Bodies of Nature

edited by
Phil Macnaghten
John Urry

SAGE Publications
London • Thousand Oaks • New Delhi

Bodies of Nature is published as Volume 6, Numbers 3–4 of *Body & Society*

First published 2001

Published in association with *Theory, Culture & Society*, Nottingham Trent University

SAGE Publications Ltd
6 Bonhill Street
London EC2A 4PU

SAGE Publications Inc
2455 Teller Road
Thousand Oaks, California 91320

SAGE Publications India Pvt Ltd
32, M-Block Market
Greater Kailash - I
New Delhi 110 048

British Library Cataloguing in Publication data

A catalogue record for this book is available from the British Library

ISBN 0-7619-7335-4

Library of Congress Control Number available

Typeset by Type Study, Scarborough, UK
Printed in Great Britain by Alden Press, Oxford

Contents

Body & Society

Books in the series:

Cyberspace/Cyberbodies/Cyberpunk
Cultures of Technological Embodiment
Edited by Mike Featherstone and Roger Burrows

Body Modification
Edited by Mike Featherstone

Bodies of Nature
Edited by Phil Macnaghten and John Urry

Bodies of Nature: Introduction

PHIL MACNAGHTEN AND JOHN URRY

... the body comes to life when coping with difficulty. (Sennett, 1994: 310)

This collection has been assembled out of diverse materials that both reflect and hopefully extend the interest in the embodied nature of people's experiences in, and of, the physical world. It thus seeks to develop further the emergent sociology of the body that has provided extensive insight into the embodied character of human experience. But such a sociology has dealt less systematically with the various social practices that are involved in being in, or passing through, nature, the countryside, the outdoors, landscape or wilderness (although see Shilling, 1993: ch. 6, drawing on Bourdieu, 1984).

These practices reflect the apparently enhanced 'culture of nature' in many contemporary societies: a culture that emphasizes valuing the natural, purchasing natural products, employing images of nature in marketing, supporting organizations concerned with conserving nature, being in the natural environment and engaging in practices that enhance the 'naturalness of one's body' (see Wilson, 1992). These practices then are part of the widely noted appeal of the 'natural', where in a way the 'social/cultural' has intervened so as to 'save' nature (see Strathern, 1992). In particular, we are concerned in this issue with the varied embodied performances implied by such practices that produce and reproduce different 'natures' (see Macnaghten and Urry, 1998, on various 'contested natures'). What bodies are necessitated by, and are reproduced through, the diverse social practices and 'contested natures' that happen within, or from, people passing through, or are occasioned by discourses of places that are in some sense in 'nature'? Why is the body, and its physical capital, developed and sustained by practices thought to be beneficial because of the 'natural' setting for such practices? How do these practices 'in nature' come to be part of the reflexivity about the body, as the self and identity are increasingly matters of deliberation, negotiation and self-monitoring (see Giddens, 1991)?

Although the embodied activities described in this volume are varied, they share some characteristics. These practices, mainly drawn from Britain, are normally constructed through discourses of 'leisure' and 'relaxation', and involve different spatialities and temporalities from everyday work and household relationships. They also depend upon bodies that have been less obviously already disciplined by work practices and relationships than would have been the case in previous historical eras in 'the West'. These 'natural' practices rely in part on post-Protestant ethic bodies. They happen in the 'outdoors', in the fresh air, where there is something about hot or cold or wet or dry 'air' that is thought particularly bracing or refreshing or rejuvenating. Such fresh air drives the body to do things or go to extremes that singularly contrast with some aspects of everyday life. Importantly, though, some of these practices become so central to people's lives that they in turn become their 'everyday', that is, when and where people actually feel really 'at home' on a wet hill-top (see Edensor, this volume), up a sheer rock-face (see Lewis, this volume), on a nudist beach (see Bell and Holliday, this volume), when wandering though a dense wood (see Macnaghten and Urry, this volume) and so on.

Most of these practices also occur beyond or in opposition to the 'urban' and the 'urban way of life', although Clark (this volume) shows the biological 'complexity' of dwelling in the contemporary city. Mostly these practices rely upon conceptions, discourses and spaces of 'nature' deemed to oppose or contradict the modernity of industry, science, the city and so on. However, such practices themselves depend upon various objects and 'mundane technologies', of boots, boats, vans, cars, bolts, towels, ropes, bicycles, sledges, spades, compasses and so on. These objects sensuously extend 'human' capacities and they deconstruct simple dichotomies of what is natural and unnatural, what is countryside and what is urban, and what are subjects and what are supposedly objects.

These practices are irreducibly embodied. But there are contradictory aspects of such a leisured embodiment. On the one hand, 'bodies in nature' are pushed to do very unusual things, to go to peripheral spaces, to place themselves in marginal situations, to exert themselves in exceptional ways, to undergo peak experiences, or to use a concatenation of the senses beyond the normal. In that sense these practices involve resistant bodies, bodies that, through their movement or their senses or their clothing (or lack of clothing) or their appropriation of times or spaces, are used in a set of resistant performances. And on the other hand, such bodies in nature are often subject to extensive forms of regimentation, monitoring and disciplining. This is realized by regulatory organizations, expert systems, technologies, moral guardians; thus, for example, much of the English countryside is deemed appropriate only for 'quiet recreation' and hence seeks to exclude those practices and people who are noisy and collective in their leisure practices (unless they are hunting foxes!). More generally, many of these resistant practices are subject to civilizing processes, of self-monitoring of the emotions, and this has made them definitive of good manners (see Shilling, 1993: ch. 7).

Bodies in nature are thus subject to novel, complex and contradictory opportunities both of escape, freedom and 'bodily naturalness' and of being constrained by modes of bodily surveillance, regulation and monitoring. The various articles show how these bodily opportunities and bodily constraints are temporally and spatially organized and managed. The practices interrogated include rock-climbing (Lewis, this volume), sailing (Matless, this volume), skiing (Ingold and Kurttila, this volume), contemplation and bodily therapy (Thrift, this volume), naturism (Bell and Holliday, this volume), being out in all weathers (see Ingold and Kurttila, this volume), walking in various contexts (see Edensor, Michael, Macnaghten and Urry, all this volume) and outdoor sex (see Bell and Holliday, this volume). We will now consider what is meant by 'nature' and the 'natural', terms that play a central role in the construction and elaboration of the practices considered here. Following that we outline some of the theoretical resources to be deployed to decipher such practices.

We can note first, and most obviously, that what has been regarded as nature has varied hugely over time and across different societies, depending in part on the notion of God/society with which it has been contrasted (see Macnaghten and Urry, 1998: ch. 1). In particular, it has been the abstraction of a 'singular nature' from the multiplicity of lived experiences that proved critical for subsequent human responses to the physical world. First as goddess, then as a divine mother, an absolute monarch, a minister, a constitutional lawyer and, finally, as a selective breeder, the appeal to a singular nature defined the changing and contested relationships between nature, God and humanity. In medieval Europe nature was viewed as a singular entity, having 'her' own place in the grand scheme of things, the Great Chain of Being. Nature was seen as God's creation, reflecting the divine and perfect order in which everything had its rightful place. God revealed himself through two books, the Bible and the book of nature. Even later, as God began to be seen as above nature and looking down on his creation, there was still presumed to be a single nature, albeit one that, from the 18th century onwards, was taken to be separate from and subordinate to society. The doctrine of human exceptionalism presupposed the belief that human progress could be measured in terms of the domination of nature, a domination that stems from viewing nature as singular, as separate from human experience and as subject to various technologies of mastery. Peoples living in or close to nature were seen as primitive and uncivilized.

From the 18th century onwards in 'the West', nature sustained 'her' separation from 'society' by departing from the predominantly human sphere to the spatial margins of industrial society.

> Nature in any other sense than that of the improvers indeed fled to the margins: to the remote, the inaccessible, the relatively barren areas. Nature was where industry was not, and then in that real but limited sense had very little to say about the operations on nature that were proceeding elsewhere. (Williams, 1972: 159)

For example, one feature of nature celebrated in Britain has been the wild, untamed and immense quality of the sea. It appears as unmediated 'nature' directly sensed but right at the margins of the land, well away from cities and civilization (Corbin, 1992). While other societies in the 'West' view as 'nature' rather different physical features: Alpine altitude and air in Switzerland, fjords in Norway, bogs in Ireland, wilderness in the US, heaths in Denmark, geysers in New Zealand and so on (Lowenthal, 1994; Schama, 1995; see Macnaghten and Urry in this volume on how the British view deciduous woods as 'natural').

We show in *Contested Natures* that rather than there being a singular nature there are various natures, which differ from, and often contradict, each other (Macnaghten and Urry, 1998). These different natures, we demonstrate, are produced by and through different social practices. There is in a way no asocial (or natural) nature, no unmediated 'pure relation' between the human body and nature (see Michael, this volume), from which unambiguously benign values and appropriate actions to save such nature can be derived. Some examples in 'the West' include nature as sets of scientific laws established from the 17th century onwards; as the open pastoral countryside which, through rural enclosures, clearances and depopulation, is exploited for upper-class field sports; as wilderness located away from industry and the cities and providing an expressive domain of purity, moral power and spiritual 'romantic' refreshment; as visual spectacle sensed through the eye and extensions of the eye (sketches, *camera obscura*, landscape paintings, postcards, photographs, camcorder); as a state of pre-social abundance and goodness reflected in the notion of 'natural' healing; and as a holistic ecosystem with a set of exhaustible resources undergoing 'global environmental change' rather than separate and localized changes.

The social practices which generate these very different natures vary in terms of how people 'dwell' within different places, especially through various 'taskscapes' of agriculture, science, industry, leisure services and so on (see Ingold, 1993, on taskscapes). They also vary as to how places are sensed through sight, smell, hearing and touch (see Rodaway, 1994), and how those senses are extended and elaborated through various mundane technologies (see various articles, this volume). Social practices also depend upon how people move in, across and beyond places through diverse mobilities. Lineado (1996) has described how the experience of the countryside and motoring are irreducibly intertwined for most people. In this collection Ingold and Kurttila focus on how changing forms of travelling over snow transform the sense and memory of weather (see also Edensor and Michael, both this volume, on mobilities of walking). Many such practices establish and sustain places of the 'other', contrasting 'unnatural' places, environments and unnatural practices that ideologically sustain the special and unique character of the bodily practice in question. Lewis (this volume), for example, describes 'sports' climbing with bolts and other equipment compared with the pure bodily 'adventure' climbing; while Matless (this volume) explores accounts extolling the virtues of the mental and physical activity of sailing as contrasted with the inactivity and laziness of motor cruising.

Many of the social practices described in this volume depend on the 'romantic' construction of nature forged through the contrasts with 19th-century industrial cities and their sense-scapes (see Lewis on the 'romantic ethic', Michael on the sublime, and Edensor on the sensory qualities of walking in the country, all this volume). Such cities were thought to be unnaturally smelly, with overwhelming odours of death, madness and poverty (Classen et al., 1994; Corbin, 1986). The rhetoric of the 'open air' was powerful for those otherwise confined to the 'unnatural' 19th-century city, which was seen as 'unnaturally' invading human orifices. In *Hard Times* Dickens described the river in Coketown that 'ran purple with ill-smelling dye', while Ruskin described 19th-century industrial London as 'that great foul city . . . stinking – a ghastly heap of fermenting brickwork, pouring out poison at every pore' (cited in Bunce, 1994: 15). The smells of the city were central to the emergent representation of the 'natural' and purifying countryside, even though rural life was at the time rich with odours of farm animals, sewage, rotting vegetables, smoke and especially foul-smelling stagnant water (Giblett, 1996). Nevertheless, industrial towns and cities have been sensed as pathological spaces, and this helped facilitate the growth of various social practices involving other senses. What emerged were practices which deployed more 'natural' senses, of touching rock-faces while rock-climbing (Lewis, this volume), of developing competences of sailing through one's 'horny' hands (Matless, this volume), of the kinesthetics of walking (see Edensor on voluntary walking, this volume), of being touched by the sun in naturism (Bell and Holliday, this volume), and of feeling the snow under one's skis (Ingold and Kurttila, this volume). These have come to be highly valorized alternative sense-scapes to those thought characteristic of the city (but see Clark, this volume, on the dynamism, poetics and vitality of city life).

How can we characterize these social practices of 'nature' and what kinds of bodies do they presuppose? What are the ways in which people dwell in nature as they engage in these various social practices? What theories might help to characterize such embodied social practices?

We begin with the distinction between land and landscape (see Heidegger, 1993; Milton, 1993; Zimmerman, 1990). The former is the mode of dwelling where *land* is a physical, tangible resource that is ploughed, sown, grazed and built upon. It is a place of work conceived of functionally. As a tangible resource, land can be bought and sold, inherited and, especially, left to children. Such land will often be directly owned and worked by the 'farmer'. Farming work, domestic work and leisure all take place in close spatial proximity. To dwell on a farm is to participate in a pattern of life where productive and unproductive activities resonate with each other and with particular tracts of land whose history and geography will be known about in intimate detail. Human subjects are apparently united with their environment since there is a lack of distance between people and things.

Landscape, by contrast, involves appearance or look (Milton, 1993), of leisure, relaxation and visual consumption by visitors. There is nothing 'natural' about landscape. Wordsworth, for example, in 1844, described how the development of landscape was a relatively recent phenomenon (1984). Early visitors to the Alps made no reference to their beauty nor to their sublime qualities that involved astonishment and terror, the dizzying claustrophobic fear induced by height, the rapid movement of water and especially overhanging rocks and crags (see Ousby, 1990). The sublime involves a simultaneous mixture of excitement and horror (see Lewis and Michael, both this volume). Wordsworth also quoted a woman who said that nowadays everyone is 'always talking about prospects: when I was young there never [sic] a thing neamed' (1984: 188). Indeed, barns and other outbuildings had often been placed in front of houses 'however beautiful the landscape which their windows might otherwise have commanded' (Wordsworth, 1984: 188). But by the mid-19th century, in western Europe relating to the countryside as landscape had become commonplace (see Green, 1990, on the visual consumption of the area around Paris). And of course by the start of the 21st century there has been a nationalization, and more recently, a globalization of landscape. Many places throughout the world have been turned into competing landscapes, places available for all those who wish to look, while driving along the open road, walking, climbing, photographing, sitting, sailing, watching TV and so on. Places have indeed been physically and semiotically designed for landscape rather than land.

Contradictions between these modes of dwelling, of land and landscape, play themselves out in the various social practices analysed in this volume. Activities in 'nature' endeavour to establish and sustain, through discourse and practice, a form of non-cognitive dwellingness analogous to what Thrift (this volume) terms a 'bare life', a life of stillness and movement, of enchantment and mysticism, of vitalism and contemplation.

These various practices are very different from the dwellingness of land, partly because of the exceptional power of the visual sense. Even while hanging for dear life on a rock overhang, it is doubtful that the look and appearance of landscape can ever be avoided. These various leisure practices also nearly always involve travel over other places of dwellingness, to get to those almost sacred sites where one can sunbathe naked, or wander 'lonely as a cloud', or sail meanderingly, or walk through a bluebell wood and so on. These practices are located in relatively distinct and specialized 'leisure times', but also in 'leisure spaces' that are geographically and ontologically distant from patterns of work and domestic routines (Macnaghten and Urry, this volume, show though that certain local woods are especially appreciated because they are 'local'). Indeed, part of the attraction of these places of nature, where bodies can be corporeally alive, natural and rejuvenated, is that they are 'other' to work, domestic routines and everyday life (the main exception to this is Ingold and Kurttila's account, this volume, of Finnish Lapland in terms of land). They involve 'adventure', islands of life resulting from bodily arousal, from bodies in motion, finding their complex way in time and space.

Lewis (this volume) draws on Simmel's analysis of the immersed, participatory body of the 'adventurer' to develop a parallel account of the situated and physical experience of the rock-climber or fell-walker. Matless (this volume) brings out the 'improving' nature of Broadland sailing practices, predicated on a unity of mind and body, conducted in a spirit of self-reliance. While Michael (this volume) quotes Szerszynski's account of 'environmental expressivism', that there is an effort to recover 'unmediated experiences' of nature not so much as a narrative of us saving nature, but of 'nature saving us, for only if we abandon modernist notions of control and domination ... can we know what to do' (1996: 121). These social practices thus involve bodily resistance where the body physicalizes its relationship with the physical world (see Jarvis, 1997, on 18th-century walking as resistance). These are not modes of dwelling either of land or of landscape, but they do contain elements of both. Elsewhere we term such modes of dwelling in, but only partly of, nature, as leisure landscapes. In such leisure landscapes, work, leisure and domestic routines are geographically and temporally estranged from each other and the physicalities of the situated body are leisured and have nothing to do with those of land per se (see Clark et al., 1994). Somewhat different is the employment of the body in protest, or in conservation holidays, which show people putting their bodies where their beliefs are (see McKay, 1996).

Most of these practices are concerned with diverse performativities, with how bodies are not fixed and given but involve various performances especially to fold nature into, and through, the body. In particular, the contributors consider the body as a bundle of senses that encounters objects and the physical world multi-sensuously. There is a complex connection between nature as a series of bodily sensations and nature as a series of socio-cultural 'sense-scapes' mediated by discourse and language (see Rodaway, 1994). In such practices, bodies perform themselves between sensation and sense-scape, moving backwards and forwards between direct sensations of the physical world and discursively mediated sense-scapes that signify taste and distinction, ideology and meaning. Bell and Holliday (this volume) show that in pre-Hitler Germany, especially the male naked body in the countryside, but not in the city let alone on the stage, was viewed as an important part of national ideology. While Matless (this volume) stresses the 'Englishness' of water, and of sailing practices as where English boys were taught to be 'at home' with water just as they were with land.

In particular, the body senses as it moves. Lewis and Thrift (this volume) both describe how bodies are endowed with kinesthetics, the sixth sense which informs one of what the body is doing in space through the sensation of movement registered in the body's joints, muscles, tendons and so on. Especially important in that sense of movement, the 'mechanics of space' (Gil, 1998: 126), is that of touch, of the hands as on a rock-face but especially of the feet as one moves in and over nature. Part of what it is that enables this kinesthetic sense to function are various objects that help to constitute actor-networks, such objects or mundane technologies sensuously extend human capacities into and across the physical world. Ingold and Kurttila (this volume) describe how changes in transport technology, of snowmobiles, motorbikes and four-wheel crawlers, transform people's sense of place and especially their sense of weather.

Michael (this volume) describes how the modest walking boot produces, together with various other objects and humans, the hybrid or quasi-subject of the 'walker'. Other contributors describe how those in nature are not there as pure human subjects but as 'actor-networks' (see Law and Hassard, 1999). Actor-networks consist of sets of humans, objects, technologies and scripts that contingently produce durability and stability, a social order of particular leisure landscapes involving various hybrids that roam the countryside and deploy the kinesthetic sense of movement (as when walking, sailing, climbing, driving the open road and so on). But Michael also (achingly?) brings out how on occasion there can be a relationship of pain between the climbing boot and the wearer, which disrupts the material flow between body and the environment. The peak experience can be contradicted by bodily sensations that index the all-too contingent and unstable actor-network!

Crucial to the analysis of such performed bodies is the concept of 'affordances' (see Lewis, Michael, Macnaghten and Urry, all this volume). Gibson argues that we do not encounter in the environment out there a set of objective 'things' that may or may not be visually perceived (1979: ch. 8; see Costall, 1995; Michael and Still, 1992). Rather different surfaces and different objects, relative to the human organism, provide affordances. They are both objective and subjective, both part of the environment and part of the organism. Affordances stem from the reciprocity between the environment and the organism and derive, as we have seen, from how people are kinesthetically active within their world (Costall, 1995: 475). Affordances do not cause behaviour but constrain it along certain possibilities: 'there are a range of options that are implicit within a physical milieu and this implicitness is directly connected to the bodily capacities and limits of the [human] organism' (Michael, 1996: 149). Some such affordances are a path that beckons people to walk along it, a rock that provides a place to hide from the sun, a wood that is a repository of childhood memories, a flat entrance that allows unhindered wheelchair access, the lake that engulfs one with cooling water and so on. There are also resistances: the heat of the sun that prevents one climbing a mountain, a road that spoils the view of a bay, the low bridge that prevents bus tourists from visiting an up-market beauty spot and so on (Costall, 1995). Given certain past and present social relations, then, particular 'objects' in the environment afford a range of possibilities and resistances, given that various hybrids in nature are active, vital, corporeal and mobile beings which are afforded various possibilities.

Finally, the emphasis upon mobility provides a link with recent applications of notions of chaos, complexity and non-linearity to social/cultural analysis (see Byrne, 1998; Cilliers, 1998). Clark (this volume) uses such notions to interpret the resurgence of 'nature' within supposedly unnatural urban areas. Certain apparently natural species have chaotically emerged in cities. Rather than urban consumerism leading to linear increases in 'civilization' and the expulsion of nature, all sorts of new natures have emerged in the city. The populations of rats, foxes and various micro-organisms have grown exponentially. Human agents have been far from successful in socializing the flows of energy and matter. The excesses of urban consumerism have thus chaotically generated new natures through 'transhuman cosmopolitanism'.

More generally, Clark describes multiple, transversal, non-linear relations within and between bacteria, viruses and other runaway mobile hazards, relations which suggest that nature has come in from the margins and may well be wreaking its cosmopolitan revenge upon human agents. Such humans are not the only species to move rapidly, indeterminately, chaotically and 'naturally' from place to place.

Note

We are very grateful to the contributors to this volume for taking on our baton and running so well with it, revealing on occasions how their own bodies 'hang out' in nature.

References

Bourdieu, P. (1984) *Distinction*. London: Routledge.
Bunce, M. (1994) *The Countryside Ideal*. London: Routledge.
Byrne, D. (1998) *Complexity Theory and the Social Sciences*. London: Routledge.
Cilliers, P. (1998) *Complexity and Post-modernism*. London: Routledge.
Clark, G., J. Darrall, R. Grove-White, P. Macnaghten and J. Urry (1994) *Leisure Landscapes*. London: CPRE.
Classen, C., D. Howes and A. Synnott (1994) *Aroma: The Cultural History of Smell*. London: Routledge.
Corbin, A. (1986) *The Foul and the Fragrant*. Leamington Spa: Berg.
Corbin, A. (1992) *The Lure of the Sea: The Discovery of the Seaside in the Western World, 1750–1840*. Cambridge: Cambridge University Press.
Costall, A. (1995) 'Socializing Affordances', *Theory and Psychology* 5: 467–81.
Giblett, R. (1996) *Postmodern Wetlands*. Edinburgh: Edinburgh University Press.
Gibson, J. (1979) *The Ecological Approach to Visual Perception*. Boston, MA: Houghton Mifflin.
Giddens, A. (1991) *Modernity and Self-Identity*. Cambridge: Polity.
Gil, J. (1998) *Metamorphoses of the Body*. Minneapolis: University of Minnesota Press.
Green, N. (1990) *The Spectacle of Nature*. Manchester: Manchester University Press.
Heidegger, M. (1993) *Basic Writings*, edited by D. Farrell Krell. London: Routledge.
Ingold, T. (1993) 'The Temporality of the Landscape', *World Archaeology* 25: 152–74.
Jarvis, R. (1997) *Romantic Writing and Pedestrian Travel*. London: Macmillan.
Law, J. and J. Hassard (eds) (1999) *Actor Network Theory and After*. Oxford: Blackwell/Sociological Review.
Lineado, M. (1996) *Car Culture and Countryside Change*. Cirencester: The National Trust.
Lowenthal, D. (1994) 'European and English Landscapes as National Symbols', in D. Hooson (ed.) *Geography and National Identity*. Oxford: Blackwell.
McKay, G. (1996) *Senseless Acts of Beauty*. London: Verso.
Macnaghten, P. and J. Urry (1998) *Contested Natures*. London: Sage.
Michael, M. (1996) *Constructing Identities*. London: Sage.
Michael, M. and A. Still (1992) 'A Resource for Resistance: Power-Knowledge and Affordance', *Theory and Society* 21: 869–88.
Milton, K. (1993) 'Land or Landscape: Rural Planning Policy and the Symbolic Construction of the Countryside', in M. Murray and J. Greer (eds) *Rural Development in Ireland*. Aldershot: Avebury.
Ousby, I. (1990) *This Englishman's England*. Cambridge: Cambridge University Press.

Rodaway, P. (1994) *Sensuous Geographies*. London: Routledge.
Schama, S. (1995) *Landscape and Memory*. London: Harper Collins.
Sennett, R. (1994) *Flesh and Stone*. London: Faber and Faber.
Shilling, C. (1993) *The Body and Social Theory*. London: Sage.
Strathern, M. (1992) *After Nature*. Cambridge: Cambridge University Press.
Szerszynski, B. (1996) 'On Knowing What to Do: Environmentalism and the Modern Problematic', in S. Lash, B. Szerszynski and B. Wynne (eds) *Risk, Environment and Modernity*. London: Sage.
Williams, R. (1972) 'Ideas of Nature', in J. Benthall (ed.) *Ecology: The Shaping Enquiry*. London: Longman.
Wilson, A. (1992) *The Culture of Nature*. Cambridge: Blackwell.
Wordsworth, W. (1984) *The Illustrated Wordsworth's Guide to the Lakes*, edited by P. Bicknell. London: Book Club Associates. (Orig. 1844.)
Zimmerman, M. (1990) *Heidegger's Confrontation with Modernity*. Bloomington: Indiana University Press.

Phil Macnaghten is Lecturer in Culture and Environment at the Centre for the Study of Environmental Change, Lancaster University. His research interests centre on the relationship between environmental concerns and wider societal trends. Recent research has looked at public controversies around genetically modified foods, the millennium, forestry, sustainability and new politics of the environment. His publications include *Contested Natures* (with John Urry; Sage, 1998).

John Urry is Professor of Sociology at Lancaster University. His recent publications include *Economies of Signs and Space* (with Scott Lash; Sage, 1994), *Consuming Places* (Routledge, 1995), *Touring Cultures* (co-edited with Chris Rojek; Routledge, 1997), *Contested Natures* (with Phil Macnaghten; Sage, 1998) and *Sociology Beyond Societies* (Routledge, 2000).

‘Botanizing on the Asphalt’? The Complex Life of Cosmopolitan Bodies

NIGEL CLARK

> . . . the town, in both its development and its form, belongs simultaneously to biological procreation, organic evolution and aesthetic creation. It is at one and the same time an object of nature and a subject of culture; an individual and a group; reality and dream . . . (Lévi-Strauss, 1975: 124)

> . . . what are the as yet untapped possibilities of complexity? This is the problem which man poses today on this planet. (Morin, 1974: 569)

If we are to believe those conversant with the broad developments of the physical sciences, recent decades have witnessed a significant shift in focus away from systems tending toward stability and simple equilibrium, towards those characterized by instability and ongoing changes of state; a turn from linear to non-linear events; from closed systems to open systems (De Landa, 1992, 1996; Kauffman, 1995; Prigogine and Stengers, 1984). As construed by the new interdisciplinary studies of dynamical systems, ‘nature’ is no longer a realm of essences or ideal states. If it is worth using the term at all, ‘nature’ is no more than the provisional outcome of local processes, the current state attained by a universe of systems whose ultimate states will always defy prediction. As compensation for the certainties that the physical sciences once aspired to, however, the world which is now materializing is more deeply imbued with creative and self-generative properties than at any other stage of our modernity; it is a restless, turbulent, unfinished place which promises surprises in perpetuity.

Over the same span, the study of cultural and social phenomena has also reconsidered its prioritizing of monolithic and enduring structures. There has been a new interest, or rather, a resuscitation of an older interest – in ‘the ephemeral, the fugitive, the contingent’ aspects of modern life, particularly in metropolitan centres where these dimensions of socio-cultural life seem most pronounced (see Baudelaire, 1964: 13; Frisby, 1985: 1–10). The city and other meeting places are being viewed afresh as the loci of converging flows and subterranean connections, as sites where complex social interactions in milieux rich in sensory stimulation are generative of new forms, practices, rhythms. As one who was precocious in his attention to the changing face of city life, the 19th-century *flâneur* has featured prominently in this revitalization of concern with the more chaotic dimensions of cosmopolitan experience (see Tester, 1994).

Walter Benjamin's depiction of the ambulant aesthete 'botanizing on the asphalt' wonderfully evokes this sense of observing the vitality of the city, as if it were a particularly fecund and diverse domain of natural life (1973: 36). And as we shall see, he is far from alone in drawing on tropes of natural history to capture the poetics of urban life. Undoubtedly, much could be mined – in the manner of Benjamin's Parisian excavations – from later manifestations of urban 'naturalism'. But what happens when we reach beyond the metaphor and delve into the seething matter of the metropolis, into the literal stuff of chaos and complexity? What if we were to discover that the city really had a 'natural life', a material existence that was not alien to the dynamical self-organizing universe of contemporary physical science?

On the other hand, 'natural life' itself now seems to have a dubious future. In line with contemporaneous environmentalist works (see McKibben, 1990), a number of social theorists have recently announced the 'end of nature' (Beck, 1992: 81; Giddens, 1994: 77; Smith, 1984: xiv). Grounding this sombre prognosis is the claim that every corner of the earth's surface evinces at least some trace of anthropogenic activity, and that every atom or molecule is potentially, if not actually, open to human modification. Effectively, this notion of a massive and irreversible socialization of the bio-physical world justifies a broad-ranging social scientific concern with the substantive dimensions of nature, one which inevitably implies 'some displacement of the sciences from a previously monopolistic position with regard to nature' (Macnaghten and Urry, 1998: 252; see also Adam, 1993: 400). What the new dynamism of the physical sciences suggests, however, is the possibility of a countermove – a permeation of the boundary between the study of the human and the non-human that passes in the other direction.

Clearly, social scientists need to be sceptical about the more grandiose claims of some complexity theorists – particularly those who seem to aspire to a kind of unified theory of everything (see Horgan, 1998: ch. 8). Nonetheless, the general transformation of the physical sciences over the last three decades or so seems to offer openings for a degree of renaturalization or rematerialization of the social world, though it is of a kind that bears little resemblance to the essentialist and ontological naturalisms rightly targeted by deconstructive practices. As Manuel De Landa has sought to demonstrate, not only is there plenty of 'life' left in organic and inorganic nature, there is a dynamism to forms and structures previously consigned to the realm of the social that has its own unique qualities, and yet belongs to the same universe of complex and open systems revealed by today's 'hard' sciences. In this way, the city literally begins to take on a life of its own. As De Landa writes: 'urban centres and living creatures must be seen as *different* dynamical systems operating far from equilibrium, that is, traversed by more or less intense flows of matter-energy that provoke their unique metamorphoses' (1997: 28; original emphasis).

It is not only that the rhythms and patterns of human systems may resonate with their organic and inorganic counterparts. By definition, all open systems are permeated by inputs from other systems. This applies not least to cities – which generally come into being at points of confluence, and continue to develop through the reprocessing of the heterogeneous elements they attract. As De Landa reminds us, alongside the cultural artefacts that conjugate at the crossroads of civilization are the quanta of energy, the inorganic matter, the genetic materials, the micro- and macro-organisms that accompany human beings on their peregrinations (1997: part II). Viewed in this 'transhuman' sense,[1] it is not only human bodies that pass through a city composed of active matter, but matter which flows in and through bodies. We might even say that Benjamin's scenario of the *flâneur* encountering a 'world in miniature' in the hidden passageways of the metropolis is iterated in a further interiorization, as the microscopic envoys of the world at large wend their way through the arterial routes and capillaries of the larger bodies who inhabit the city (see Benjamin, 1978: 147).

Recent discourses on risk have gone some way in drawing the flow of matter and energy – and the openness of the body to these flows – into contemporary social thought. With its focus on the way that modified natural elements stray from their assigned paths into our environs, our bodies and our consciousness, Ulrich Beck's notion of undelimitable accidents has injected a novel sensitivity to bio-physical materiality into the heart of social analysis (1995). We might even say that the concept of risk society begins to credit matter with a cosmopolitanism of its own, intimating a transhuman take on Simmel's stranger who comes today and stays for endless tomorrows. And yet, in spite of his equation of globalized risk with cosmopolitan society (1996), Beck's work skirts around some vital aspects of urban and global experience. According to Scott Lash, both Beck and Anthony Giddens put their emphasis on the scientific or cognitive experience of risk at the expense of any aesthetic dimension it might have. That is, in their quest for more secure social arrangements, these two pre-eminent theorists of risk society focus on the contradictions deriving from the functioning of modern institutions in their rational and utilitarian capacity. As Lash puts it:

> ... their notions of ambivalence and experiment come largely from scientific ideas of indeterminacy, and the unintended consequences of the interventions of science. This is insufficient. In everyday life I expect that such a sensibility derives a lot more from the aesthetic or hermeneutic sensibility. (1994: 213)

With its privileging of turbulence over stability, surprise over predictability and stimulation over safety, the cosmopolitan aesthetic to which Lash draws our attention appears to pose fundamental challenges to any project of reorganizing contemporary socio-economic life that is predicated on a quest for security. Without denying that contemporary urban existence is rife with fear, paranoia and a sense of bodies under siege, we must also take account of a certain ' willingness to take risks' (Urry, 1995: 167), or what the surrealist Louis Aragon (1970: 13) describes as 'a taste for the promiscuous and the unstable', that are equally vital aspects of cosmopolitanism.

But now that 'a taste for the promiscuous and the unstable' belongs as much to science as to the arts, can we continue to counterpoint aesthetic sensibilities with rationalities derived from the realm of cognition? In other words, we must ask whether Beck and Giddens – or Lash for that matter – have made enough of contemporary 'scientific ideas of indeterminacy'. What becomes of the old Kantian categories when we extend aesthetic discourses on metropolitan existence to include the rhythm and flux of non-human elements, and at the same time expand the notion of complex, dynamical systems to take in the contingencies of human sociation and cultural life? What might we learn for the future of the embodied human agent – if we follow De Landa's lead and consider the human being as an open system – as a temporary congealing of flows of matter, energy and information through a wider world? And how might we rethink our fearful exposure to an ever-changing stream of organic and inorganic elements, if we come to perceive ourselves as always already the product of chance encounters of heterogeneous materials: entities and elements that have been subject to an extended sourcing for as long as travelling bodies have met at crossroads or converged on permanent settlements?

Botanizing on the Asphalt

I want to begin not with the science of dynamical systems, but with that appreciation of chaos and complexity which precedes the present interest by almost a century and half. Ironically, in order to return to nature – or to rematerialize culture – we will set out from an adverse beginning: an apparent turn away from the natural. In the mid-19th century, the poet and essayist Charles Baudelaire, a youthful believer in the beneficence and inherent harmony of nature, turns his back on the notion of nature as the master-image or great referent of art (1964: 31–2, 46). This move, it has often been suggested, opens the way to a vision of the city as a landscape in its own right. Not simply repudiating nature, he elevates '*le paysage des grandes villes*' to the same status, so that it too can provide 'the delights of chaos and immensity' once reserved for the natural sublime (Baudelaire cited in Leakey, 1969: 171). As the bearer of this new cosmopolitan subjectivity, Baudelaire elects the artist-as-*flâneur*: the strolling aesthete who gazes afresh at the urban spectacle, registering the changing seasons of its fashions, the subtle transformations of manner and gesture, the life and death of each new commodity.

A century or so later, Walter Benjamin probes the significance of Baudelaire's discoveries, gathering further examples of the 19th-century reinterpretation of city life, and exploring the metaphor of the astute urban spectator as a new form of 'naturalist'. Benjamin draws our attention to the journalistic descriptions or 'physiologies' of city life, popular in the early 1840s, equating the genre with the 'style of the *flâneur* who goes botanizing on the asphalt' (1973: 35–6). It is not only Baudelaire, he notes, but a host of literary figures who appropriate the tropes of the writing of the wilderness and frontier to evoke the unruliness of urban existence. Just as Baudelaire exclaims: 'What are the dangers of the forest and the prairie compared with the daily shocks and conflicts of civilization?', so Alexander Dumas opens *Mohicans de Paris* in 'The tropical forest in the Rue d'Enfer', while Victor Hugo observes of an inner-city neighbourhood: 'What had happened on this street would not have astonished a forest' (cited in Benjamin, 1973: 39, 41, 62). As later commentators have noted, the naturalist metaphor is also enlisted in Proust's *A la recherche du temps perdu*, when the narrator Marcel is compared to a botanist, as he observes an encounter between two previously unacquainted homosexuals and likens it to the chance fertilization of a rare orchid by a passing bee (Wilson, 1995: 65). And likewise in Joyce's *Ulysses*, where the 'submerged' sector of the urban society is subjected to a turn 'under the microscope' (Lechte, 1995: 104).

To its early and high modernist observers, then, the city is the site where culture or civilization has grown dense, heterogeneous and mutable. Fixed categories and inherited distinctions no longer render social space intelligible, the shifting appearances of the city suggesting turbulence beneath the surface that can be registered only by a vigilant gaze. But here we reach the limits of metaphors derived from the empirical sciences, for this cosmopolitan aestheticism deploys its rigorous optic not so much to open the cultural universe to the imperatives of order and clarity, as to affirm its unregenerate complexity and disorder. If nature

– near or distant – was destined to fall under the sway of the ordering gaze, the city would keep making itself afresh, generating ever new forms and configurations. Where scientific explorers of new regions could establish their bearings in the wildest of seas and jungles, wandering aesthetes could with comparable effort disorientate themselves in the most familiar environs. 'Not to find one's way in a city may well be uninteresting and banal', wrote Benjamin. 'But to lose oneself in a city – as one *loses oneself in a forest* – that calls for quite a different schooling' (Benjamin, cited in Buci-Glucksmann, 1994: 84).

It was this sense of wilful derangement that Benjamin admired so much in Louis Aragon's *Le Paysan de Paris*: a disjointed account of urban excursions in which the narrator infuses the city with mythic or auratic qualities fitting to the modern era (Buck-Morss, 1989: 256–7). To be attuned to the ephemera and contingencies of the urban matrix, then, was to recognize that it exceeded reason; it was to experience the world in a way that – at least momentarily – escaped the strictures and statutes of bourgeois authority. But this idea of another reality, an underworld with its own logic, is by no means unique to urban-industrial modernity. As Benjamin put it: 'the city is the realization of the ancient dream of the labyrinth. Without knowing it, the *flâneur* goes in search of this reality' (cited in Frisby, 1985: 94).

Baroque Reason: The Logic of Fragments

It was in his early study of 17th-century Baroque tragic drama or *Trauerspiel* that Benjamin first tracked the way that a system of meaning and order could generate its own excesses, so that from the heart of a seemingly orderly arrangement of signs and things could come 'a veritable eruption of images, which gives rise to a chaotic mass of metaphors' (1977: 173). In the industrializing metropolis of the 19th century, Benjamin recognized that the proliferation of signifying objects again exceeded the capacity of received narratives for ordering and containment – opening the possibility for alternative readings of the urban milieu. More recently, perhaps from the midst of another surge of excessive signification, a number of theorists have observed that a 'baroque' or 'allegorical' sensibility seems to shadow more orderly and linear conceptions of history, surfacing again and again over the course of our modernity (Buci-Glucksmann, 1994; Jay, 1992; Lash, 1993, 1994; Maffesoli, 1996; Stafford, 1996).

This sensibility – this 'baroque reason', as Christine Buci-Glucksmann terms it, is the 'uncanny' of modernity, the insistent reminder that materializes in art, writing and certain urban experiences that reality has become 'enigmatic, hieroglyphic, non-rational' (1994: 45). Where the dominant mode of Western reason sets itself to illuminating the darkness and bringing the hidden depths into the clear light of day, 'baroque reason' assumes that occlusion, secrecy and subterfuge will always be with us. Where classical or Cartesian reason places its emphasis on a chain of causes and effects compounding to a linear sense of progress, the baroque sensibility points to the promiscuous relation of causes and effects, to indeterminacy and unclear direction. And where the classical *ratio* aspires to totality, the baroque elevates the fragment in all its limitless conjunctions with other fragments. 'Just as the classical is linear, visual, closed, analytical, and liable to be clearly analysed', Michel Maffesoli observes, 'the baroque is evolving, complicated, open, synthetic and evokes a relative obscurity' (1996: 159).

Unlike the Romantic opposition to reason, the baroque impulse does not offer an alternative vision of the whole, eschewing the counterpointing of an ideal world to the fallen present. For modern allegorists – following Baudelaire – the prevailing experience of modernity is perhaps that of ambivalence: in lieu of a simple valorization of the new and the progressive there is pleasure to be had from the unexpected, the random, the ephemeral – but this is tempered by the possibility of catastrophe and the omnipresence of ruin and decay (Buci-Glucksmann, 1994: 102). And nowhere is this ambivalence inscribed more deeply than on the human body.

If classical reason privileges human cerebral qualities, and values the body primarily as a symbol of universal proportion and ideal beauty, the baroque sensibility construes the body as a site of impurity and corruptibility, at once the locus of pleasure and of suffering, of the most beautiful seductions and of the basest deceptions. For the male allegorist, encountering the increasingly commodified libidinal economies of the 19th-century city, it is women's bodies that most obviously bear the signs of a plummet from the classical pedestal. And it is the figure of the prostitute who epitomizes this shattering of the ideal and paradoxical feminine potential for both dangerous and pleasurable duplicity that is its outcome. As Buci-Glucksmann claims, 'the prostituted body is not only fragment, ruin of nature, disfigurement of the "sublime body". It is also a staging in and through new imaginaries created by a thousand excitations' (1994: 102).

As has often been noted, male connoisseurs of the new human nature of the metropolis – the naturalists of 'female fauna' and all the other facets of the urban phantasmagoria – paid a price for their optical perspicacity. In order to observe without being taken in, to register the recalibrations of the commodity, the crowd and the cityscape without succumbing to their charms, the *flâneur* assumed a stance of detachment – or permitted himself only the briefest of dalliances. Which is to say, for all his refined aesthetic or cultural sensibilities, he remained socially disengaged (Ferguson, 1994: 26, 38; Shields, 1994: 77; Wilson, 1995: 73–4). As

Benjamin notes unceremoniously, Baudelaire 'throws his lot in with the asocial. His only sexual communion is realized with a whore' (1978: 157).

The Sociability of the Underworld

Scott Lash's work suggests that the atomized life of the *flâneur* is symptomatic of a more general problem with cosmopolitan aestheticism as a countercurrent to rational or cognitive constructions of the modern (1994: 135–53). In its privileging of the particular over the universal, he argues, the allegorical tradition remains fixated on individual experience. Whether we are dealing with Baudelaire, Nietzsche or today's deconstructivists, the affirmation of heterogeneous desires suffers from a failure to identify the communal dimension – the shared meanings and practices – which underpin such experiences (Lash, 1994: 147).

We gain a different perspective on the poetics of cosmopolitanism, however, if we follow Maffesoli's lead, and conceive of meaning, empathy or communality not so much as the prerequisites but as the *emergent properties* of unscripted urban encounters. In an inversion of the top–down approach more typical of sociological analysis – the question of what structures or institutions must be in place in order that social action and cultural processes occur – Maffesoli engages with sociation from the bottom up – asking, rather, how patterned behaviour and shared meaning might congeal from the everyday chaos of human conjugation. From this angle, it is not so much the activities of the *flâneur* that we need to follow through, but the encounters to which he bears witness: not Proust's narrator as biologist, but the warm bodies who meet and bond on the street before him, whether he is watching or not.

In *The Time of the Tribes* (Maffesoli, 1996) it is the urban 'masses' themselves – going about their everyday lives – who are credited with creativity. Their mode of aesthetic expression *is* the construction of communal forms: by forging bonds, creating their own micro-groups or 'neo-tribes' and developing new lifestyles they construct something that is more than the sum of its parts (Maffesoli, 1996: 96). Theirs is a 'baroque sociality', a subterranean proclivity to form associations that asserts itself in any dense and complex agglomeration of human beings, though more conspicuously 'at times of fracturing, disturbance or effervescence' (1996: 159).

Lash is doubtful whether new conceptions of ethical-aesthetic bonding – like Maffesoli's neo-tribes – fit the criteria of full and authentic 'community': he wonders if they are 'but merely associations of atomized individuals' (1994: 143–4). But like the shattering of essences, the dispersal of ideal forms that frees the fragment to seek out its own destiny in the baroque vision, Maffesoli's bottom–up conception of sociation disavows the defining traits of true community, crediting each association with its own contingent outcome. Only a posteriori can we gain a sense of the forms and structures that emerge from the tangle of social exchanges constitutive of complex societies: at the concrete, lived level we know only what is local or proximate, only what is adjacent in the networks which we create and which recreate us. As Maffesoli argues, Simmel grasped the inherent opacity of the social life of the metropolis when he described the social process as operating through a 'multitude of minuscule canals, the existence of which is unknown to individual consciousness' (cited in Maffesoli, 1996: 89). In this sense, the processes of sociation do not require the constant reflexivity that is so important to Lash, or in a different mode to Anthony Giddens or Ulrich Beck. Intentionality plays its part, but the plethora of small acts which bring us together are not necessarily logical nor even conscious. As Maffesoli concludes: 'we should not be afraid to say it – we lead what is quite an animalistic life' (1996: 88–90).

By transfiguring the *flâneur* into a social being: one who follows up on the chance encounter, who forges crossed paths into networks of communication, reciprocity and sensual exchange, Maffesoli grants the abstract desires of cosmopolitan existence a material body. Viewed in this way, the cosmopolitan is no longer a species apart, an exile from the natural order. For Maffesoli, 'the vitality of the masses' – this spontaneous intermeshing of warm, emotive human bodies into group formations – is a natural process; no less so than the bonding of chemicals or the accretion of coral (1996: 51, 34–6). No longer merely the object of a metaphorical botanizing, the secret life of cities has materialized into an organic system in its own right: a manifestion of the sort of fecund, effusive energy that flowed continuously through both natural and social domains in the world according to Georges Bataille (Bataille, 1991; Maffesoli, 1996: 51, 107).

Chaos, Complexity and Form

If the baroque sensibility is born of the death of one nature: a world of essences, ideal forms, timeless verities – so too does it announce a new nature: one of unpredictable transmutations, fusions, irruptions, disappearances. This is a nature in which no templates have been provided, where parts do not gain meaning from a pre-given whole, nor the present from an anticipated future. Baroque reason, as outlined by Buci-Glucksmann, speaks of an underworld beyond the usual limits of Western philosophy which is the hidden breeding ground of new forms and possibilities. 'Not only does beauty rise up from the abyss; there is no form except on the basis of a primal, chaotic, maternal formlessness, which explodes the classical Hegelian hierarchy of form and content' (Buci-Glucksmann, 1994: 151).

Social science, it has been argued of late, has a general tendency to follow the dominant traditions of Western philosophy in its allegiance to analytic procedures. Though there are exceptions, this means that it is prone to start with the finished product of social action, which in turn tends to imply thinking in terms of a teleological movement toward some ideal or steady state (De Landa, 1997: 270, 1993: 797). The price to be paid for such a 'top–down' approach is a lack of attention to the emergence of systematicity itself – a bypassing of 'the *specific* and *contingent* histories' by which forms and structures are constantly coming into being – or dematerializing (Featherstone and Burrows, 1995: 14–15; see also Cilliers, 1998: 89; De Landa, 1996: 183). It is in this sense that the interdisciplinary study of complex systems might usefully complement the resurgent interest in a baroque or allegorical 'other modernity', for both encapsulate a turn away from an analytic toward a bottom–up or synthetic approach: the shift 'from the model to the matrix' which Benjamin long ago anticipated (cited in Buci-Glucksmann, 1994: 59). We might see Gilles Deleuze's interest in both the complex enfoldings characteristic of baroque art and thought, and the turn of late 20th-century science toward open systems and non-linearity as a vital prefiguring of any such conjunction (1993, 1995: 31–2).

Reminiscent of the generative and chaotic abyss of Buci-Glucksmann's baroque, studies of complexity deal with 'the shadowy zone . . . the obscure nucleus of self-organizational logic where everything seems to us to be mixed up in what is inexpressible and contradictory' (Morin, 1974: 573). It is here that new forms come into being, in these regions that we encounter 'the alchemical process in which . . . transmutation takes place' (Maffesoli, 1996: 20). Or what we might more technically call 'morphogenesis' – the spontaneous emergence of form and order out of more amorphous elements.

Such processes of form-building or self-organization are central to complexity studies, which take as their object any system composed of enough locally connected parts to process inputs from its environment in an adaptive or generative manner. As complexity theory would have it, when matter, energy or information flows into such a system, it passes from node to node: the density of interconnections and the multitude of possible combinations ensuring the pathway taken cannot be determined in advance. When these flows are repeated often enough, the system is capable of 'remembering' certain pathways, which is to say storing and reselecting certain patterns. This compounded learning can bring about a shift to a new level of organization or complexity, a reconfiguration of parts into a new whole with properties which cannot be derived from the elements viewed in isolation (Cilliers, 1998: ch. 6). Alternatively, novel inputs – amplified by their resounding through the nodes and networks of the system –

can lead to a sudden loss of order, a crash or catastrophe (Prigogine and Stengers, 1984: 206).

A self-organizing or dynamic system, then, has a history. While there may be common patterns of emergence across a range of complex systems, the actual outcome of any event is contingent on the interplay of inputs with the elements internal to the system. As ever more of the universe is revealed to be constituted not of closed systems, nor of simple, predictable linear systems – but of systems characterized by openness, self-transformation and an inherent unpredictability, it is the so-called 'hard' sciences which are emerging as prime advocates of open-ended becoming. Moreover, the impossibility of representing the pure immanence of self-organization with anything less than simulations of equal complexity is putting the science of complex dynamical systems firmly on the side of the critics of classical models of representation. 'Meaning is conferred not by one-to-one correspondence of a symbol with some external concept or object', Paul Cilliers observes, 'but by the relationships between the structural components of the system itself' (1998: 11, ch. 5).

In the study of self-organizing systems, surprise seems to be valued above all else: a system 'poised at the edge of chaos' is deemed the most fruitful site of study, the emergence of something from nothing the hoped-for occurrence (De Landa, 1992: 159). In the case of the computerized simulation of a self-organizing system, '[t]he exercise will be considered successful if novel properties, unimagined by the designer, emerge spontaneously' (De Landa, 1993: 800). In this sense, the ascendant paradigms associated with study of emergent behaviour or morphogenesis deliver a new challenge to Kantian distinctions between aesthetic and cognitive value spheres. No longer, it seems, can the subject of cognition be unproblematically equated with calculating ambitions and the quest for certainty, nor simply set apart from an aesthetic subjectivity siding with chance, incalculability or fugitive effects. No more does the aesthetic critique of modernity have a privileged relationship with, in Lash's terms ' the language of "difference", ... "complexity", "excess", "flux", "contingency"' (1993: 14).

As it now stands the tracking of complex dynamical systems plays across the distinctions between the organic and inorganic, the living and the non-living, the human and the non-human: the burgeoning interdisciplinary field being as likely to concern itself with the flow of traffic or the networking of human actors in a city as it is with turbulent meteorological phenomena, the relationship between species in an ecosystem, or the functioning of neurones in the brain. Partially recalling the 'urban ecologies' of the Chicago School, though relieved of their tendency to posit a climax state as the outcome of socio-spatial succession (see Hannerz, 1980: ch. 2), the new paradigm opens the way to a conception of the city as an open system; as a filtering and sorting device with an inherent capacity for self-organization.

From 'the daily metamorphosis of external things' in Baudelaire's Paris (1964: 4) to the 'rhythmic throb and the perpetual discord and dislocation of all opposing rhythms' of Robert Musil's Vienna (1953: 11), the great aesthetic renditions of the urban dynamic take on a new cast in the light of recent science.[2] In this context we might view the urban aesthete – with his or her attraction to flux and indeterminacy – as a kind of precursive connoisseur of the generative powers of complex dynamics. Amidst the ruins of determinism and linear thought, what correspondences might now open up between the motifs of the baroque sensibility – the choria, the labyrinth, the infinite library – and the technoscientific tropes of non-linear relationality: the matrix, the rhizome, the neural net and the dissipative structure?

The Transhuman City

Countermanding two centuries of romantic revolt against the unnatural existence spawned by urban civilization, Maffesoli chooses the depths of the metropolis as the site of his assault on 'the arbitrary division . . . between nature and the sciences of mind' (1996: 36). And yet, for all that the vital energy of the universe pulses through the boulevards and by-ways of his cities, its flow confines itself to fields composed of human forms and artifacts. The body of his urban dweller may open itself to the warmth of other human bodies, but its porosity to the non human entities and elements that pass through the city remains to be explored. In short, where an older baroque aesthetic has contemplated the truly monstrous and chimerical, Maffesoli's sociological version limits itself to socio-cultural transmutations.

Urbanized human beings, we may not need reminding, have historically been less than hospitable to most of the organisms they share their cities with. As De Landa points out, the main biological characteristic of the urban ecosystem is the shortening of food chains, at the expense of species seen to compete with *Homo sapiens* for nutrients or space (1997: 108). A process of banishment, it might be added, that is echoed in the anthropocentric circumscribing of the social sciences, the humanities and most other narrations of Western civilization. But, as De Landa is quick to add, the biotic homogenization of cities has always been compromised, with some non-human species making regular or periodic returns (1997: 109).

On the one hand, urban dwellers have gone to some lengths to compensate for the thinning of non-human life in their presence by deliberately reintroducing species in captive or domesticated modes. Just as cities have acted as gateways to foreign cultures, so too have they been crossroads for the movement of exotic biota. The assembling of a world of biodiversity has been institutionalized in the West in a succession of forms – including circuses, botanic gardens, parks and menageries, natural history museums, zoos, aviaries and aquariums – whose location in the heart of the metropolis prompts a further literalization of 'botanizing on the asphalt'. For Benjamin, such intermingling of 'old nature' with the new nature of urban industrial forms – exemplified by the enclosure of elms, palms and other greenery in the Crystal Palace in the 1851 World Fair – was integral to his vision of the modern city's realization of ancient myths of harmony and plenitude (Buck-Morss, 1989: 110–14). Perhaps as a more generalized manifestation of the same desires, most modern cities have been home to a vigorous planting of both domestic and exotic species, resulting in what Anne Whiston Spirn describes as a distinctly 'cosmopolitan flora' (1984: 26).

But what is more challenging to our received conceptions of the city and the bodies therein, and our writing of civilization in general, are the undesigned and undesired resurgences of nature in urban spaces. For just as 'the fauna and aquatic plantlife of the imagination huddle and survive in the ill-lit zones of human activity' in Louis Aragon's literary vision of the metropolis (1970: 9), so too have their material counterparts persisted, and often thrived in the netherworlds of socially organized space. As De Landa has so eloquently underscored, human agents have been far from universally successful in their 'socializing' of the flows of energy and matter (1997: 122). Cities, he reminds us, are dynamic and open systems, the multiple forms of matter-energy (including minerals, biomass and genes) which pass through them entering into complex, non-linear relationships whose outcomes tend to exceed the calculations of their human component (1997: part II).

Indeed, viewing our civilizational achievements from the point of view of their other, normally overlooked participants offers a strong counterforce to those narratives which assume steady human ascendance over lesser beings. As Deleuze and Guattari propose: 'the apparent order can be reversed, with cultural or technical phenomena providing a fertile soil, a good soup, for the development of insects, bacteria, germs, or even particles' (1987: 69). Across the built-up areas of the globe, certain species – hardy, adaptable generalists across a range of phyla – have successfully (re)established themselves, and proved extremely difficult to dislodge (Quammen, 1998: 67; Spirn, 1984: 207). Such 'weedy species' – including crows and cockroaches, rats and ragweed – have taken advantage of human vectors of transportation, they have capitalized on the removal of potential competitor species and have accustomed themselves to the diverse offerings of urban waste. Rat populations, it has been estimated, exceed human numbers in the average city (Spirn, 1984: 207), as would many smaller colonists.

In his recent fusion of dynamical systems theory and urban history, Mike Davis (1998) enlists the cataclysmic evidence of southern California behind the argument that the 'wild' and the 'urban' are far from exclusive categories. Along with the recalcitrance of floods, fires and earthquakes, he tracks the unexpected resurgence of wildlife along the fringes of urban settlement. Davis draws attention to the innovative behaviour of fauna along the 'increasingly fractal edge of development' (1998: 237, ch. 5) – what would seem to be an instance of the 'edge-of-chaos' unpredictability beloved of complexity theorists. Joining coyotes, racoons and other known generalist species, he argues that mountain lions may now be in the process of a critical transition from highly specialist predation to a more broad-based opportunism – as they learn to feed off pets, small wildlife lured by garbage, and the occasional human camper or jogger. 'What we are witnessing', Davis claims with scarcely suppressed glee, 'may be nothing less than a behavioural quantum jump: the emergence of nonlinear lions with a lusty appetite for slow, soft animals in spandex' (1998: 249).

Baroque Life: The Micro-Organismic Underworld

With junk food leftovers, pets fattened on gourmet cat-food and even the conveniently encaged fauna of the city zoo featuring in the new food chains of the southern Californian conurbation, it would appear that the excesses of urban consumer culture are spawning an ecological-era monstrousness. We might be forgiven for thinking that Davis's feline predators are inheriting the mantle of the Victorian prostitute as the devourers of unwary cosmopolitan bodies and souls. But perhaps where the most truly baroque emergent properties of the contemporary metropolis are to be found – where the most devious transformations and the most promiscuous exchanges take place – is deep in the underbelly of the social, below the horizon of the unaided human eye.

In the day-to-day realm of human inhabitation of the metropolis, the secret life of cities that has proved the most unnerving is the activity of micro-organisms – the unanticipated events that take place not only around us, but within us. As we know from the work of Foucault, fear of plagues played a major role in the administrative regulation of the early modern city (1991: 144). Ironically, the same efflorescence of organismic life in the cities of Europe seems to have been a precondition for the aesthetic appreciation of untrammelled nature. 'After the experience of the great urban epidemics', Klaus Eder observes, 'the country became sentimentally the locus of a better life, a life closer to nature' (1996: 177).

It is hardly surprising that organisms which harvest human biomass from within, often fatally, should provoke widespread horror and loathing. Like other complex, non-linear systems, the relationship between contagious 'micropredators' and their hosts is an inherently unpredictable one: though most will eventually pass through the phases of epidemicity and endemicity on the way to some form of mutual accommodation (De Landa, 1997: 109). Drawing on pioneering work by William McNeill and Alfred Crosby, two general historians who have credited micro-organisms with a certain socio-historical agency of their own, De Landa traces the emergence of new diseases to the densely packed consociation of human and domestic animals that comprised ancient and medieval cities (1997: 110, 123). In the centuries which followed, the opening up of new communication channels between centres served to unify distinct disease pools across Eurasia during the 14th century, reaching out to encompass most of the rest of the globe by the mid 18th century (Arnold, 1996: 74–7; De Landa, 1997: 123–34).

After initially cataclysmic epidemic phases, 'global unification by disease' has generally settled into endemicity or accommodation, aided by hygienic and immunological programmes, and by the development of antibiotics in the mid 20th century. But, as De Landa concludes, 'the fact that modern medicine has gained a larger measure of control over micro-organisms does not mean that we have ceased to form a meshwork with bacteria, viruses, plasmodia, fungi and other "weeds"' (1997: 174). Contrary to expectations, the liberal deployment of penicillin and other antibiotics has met with highly effectual counter-strategies on the part of the target organisms (Van Loon, 1997). It is not only that surviving bacteria spawn resistant lineages, they have also taken advantage of the transversal exchange of genetic material between micro-organisms to transmit antibiotic resistance across strains, even across species (De Landa, 1997: 174–5).

As microbiologist Lynn Margulis has emphasized, bacteria do not respect the hereditary regulations or boundaries established by scientists more familiar with the reproductive behaviour of organisms close to their own size. Indeed, all bacteria may be seen as nodes in one great web of interconnectivity; swapping genetic material in all directions, irrespective of lines of descent or speciation (Margulis and Sagan, 1995: 73–6; Sagan, 1992: 376–7). Extending this reticular logic, some viruses are capable of transferring genetic material from one host organism to another, even across species boundaries – so that, as in the world of bacteria, other phyla are connected in transversal, non-linear modes (De Landa, 1997: 138). In the words of Deleuze and Guattari: 'We form a rhizome with our viruses, or rather our viruses cause us to form a rhizome with other animals' (1987: 10).

Such transgenic exchanges, biologist François Jacob observes, are reminiscent of 'the abominable couplings dear to antiquity and the Middle Ages' (cited in Deleuze and Guattari, 1987: 11). Or to that recurrent 'other modernity' we have termed the baroque. For just as Maffesoli's 'baroque sociability' is constituted of individuals whose excursions through the networks of the city effect chance meetings and novel congregations, so too do the movements of micro-organisms – now facilitated by our own networks of transportation – result in the spontaneous formation of new communities: bacterial consortia with their own emergent properties and structures. While Margulis and Sagan write of 'post-bacterial cells with erotic habits mirroring our own' (1995: 13), it is of course our cosmopolitan sexual adventures that echo a micro-organismic miscegenation billions of years our senior.

Margulis's claim that complex organisms develop through enclosure of more simple organisms is now widely accepted in both scientific and popular circles (see Margulis and Sagan, 1995: 17). On the account of such symbiotic incorporations, the human body – like any animal or plant – can be viewed as a 'seething zoo of microbes', a reliquary of a long and continuing history of ingestions and mergers (Clair Folsome, cited in Margulis and Sagan, 1995: 178). Moreover, many of these symbioses, Margulis argues, began as life-threatening infections, only to emerge as new and useful body parts (Margulis and Sagan, 1995: 107; see also Morin, 1974: 577). Such involutions are not limited to organic 'components', as the micro-organisms themselves may have been formed from more elemental incorporations, as in the case of bacteria which have ingested iron, enabling them to orient themselves towards the nearest magnetic pole (Margulis and Sagan, 1995: 52).

Fear and Fascination in the World of Dynamical Systems

Increasingly, then, the organismic body is being construed as an open system, an unfinalized product of morphogenesis, a congregation of heterogeneous materials and flows. 'Corporeal existence', Vicki Kirby writes, 'is generative and generous in its inclusiveness, an infinite partitioning, mediated from and within itself; an animated representation whose fractured mirroring includes cellular and atomic life' (1997: 146).

As Donna Haraway (1991) famously proclaimed over a decade and half ago, the human experience of interfacing with electronic information systems has contributed profoundly to such a sense of the erosion of bodily boundedness and integrity. For subsequent theorists of the condition of embodiment, however, the notion of open and composite bodies has been extruded temporally and conceptually far beyond the reaches of contemporary informatics or technoscience: with such anciently adulterated entities as gene-trading prokaryotes and iron-ingesting bacteria now appearing as the always already of cyborg ontology (Ansell Pearson, 1997; Plant, 1997).

Extending rather than undermining Haraway's seminal insights, Sadie Plant implicates the proliferation of information networks with a growing appreciation of open systems and spontaneous, unpredictable becomings. 'Of all the media and machines to have emerged in the late twentieth century', she writes, 'the net has been taken to epitomize the shape of this new distributed nonlinear world' (1997: 46). As this realm of digitized communication and data-processing grows ever more labyrinthine, an increasing proportion of tasks and operations is being ceded to autonomous programs: software constructs which may be gifted with the capacity for self-dispersal, replication, collaboration and even a degree of self-transformation or 'evolution'. There is now a whole phylum of programs which, as both technoscience commentators and social theorists have observed, are demonstrating a significant range of the properties conventionally attributed to biological life (Dibbell, 1995; cf. Luke, 1997: 14–18). And just as some of us may find a certain appeal in Mike Davis's account of the reclaiming of suburbia by predatory megafauna, many observers find a source of enchantment in the populating of the Internet with life-like entities; even, or perhaps especially, when the 'creatures' in question appear to display wilfully errant and refractory qualities (see Dibbell, 1995).

From the shadows of pronouncements of the 'end of nature' – with its melancholic and monolithic vision of the retreat of nature before socio-technological colonization – are such affirmations surprising? In the context of claims that even the most remote tracts of 'wild' nature bear the imprint of the social, there is an understandable attraction to the idea that in the very heartland of the social – be it the sprawl of suburbia or the depths of the datasphere – there is a resurgence of 'nature', an efflorescence of 'life'. As a counterforce to the unidirectionality of the end of nature theses, then, such irruptions seem to offer a welcome vindication to Deleuze and Guattari's suggestion that, under certain conditions, 'cultural or technical phenomena' provide a fertile environment for life-forms and active matter.

The idea that social, economic or technological systems might display similar rhythms and periodicities to systems in the rest of the universe adds depth to these counterpoints to the final socialization of nature. It must be added, however, that a vision of mutually implicated and resonating socio-technical and biophysical systems by no means detracts from the gravity of the environmental predicament that the end of nature theorists like Beck or Giddens wish to impress upon their audience. On the contrary, De Landa's exploration of the wider socio-cultural significance of a new insight into the transformation of matter–energy converges on the ecological crisis, and he permits himself only the most guarded optimism that recent advances 'indicate there may be ways of evading our currently doomed environmental destiny' (De Landa, 1992: 161). Furthermore, in the work of De Landa, as in the more poetic but less systematically scientific explorations of self-organizing matter–energy in the work of Deleuze and Guattari (1987: 503), there is a clear sensitivity to the risks residing in the new knowledge, an awareness that imprudent experiments with any sort of system at a critical point might as easily prompt catastrophic breakdowns as they would constructive unfoldings.

What the more circumspect commentators on the field of non-linear dynamics seem to be cautioning us about is that in learning to model self-organizing processes, though insights into the evasion of undesirable systemic disruption may be gained, we simultaneously advance our capacity to nudge systems into far from equilibrium states or to precipitate self-accelerating events. Which is to say that risk theory, as outlined by Beck (1992, 1995), and certain manifestations of non-linear or complex systems theory would seem to have some important potentials for mutual illumination. With regard to putting new knowledge about critical states of matter-energy into practice, a deep concern for the environment or for the future of society might prove as much a incentive as a deterrent: for, as Zygmunt Bauman has warned, '[c]atastrophes most horrid are born – or likely to be born – out of the war against catastrophes' (1992: 25). In other words, by attempting to confront 'out of control' events on their own terrain, we may be drawn toward solutions that mirror the problem in their dispersability and intractability, and hence in their catastrophic consequences (Clark, 1998).

It is in this context that we need to take full cognizance of the aesthetic dimension of the growing interest in unpredictable morphogeneses, 'systems on the edge of chaos' and runaway technological events. For where Beck (1995: 109) construes 'an "open-ended festival" of creeping, galloping and overlapping despoliation' as a terrifying prospect, a number of other commentators on the breakthrough to 'self-accelerating and self-proliferating' technology appear more sanguine, cheerfully anticipating a 'carnival ... of constant innovation and surprise, with desired results and unexpected side effects colliding in all directions' (Brand, 1991: 6; see also Kelly, 1994: 401).

While many critical social and cultural theorists remain firmly focused on the dominating and controlling imperative of modern science, I have been suggesting that contemporary technoscience is no stranger to that 'taste for the promiscuous and the unstable' that has long been axiomatic to cosmopolitan aestheticism. Indeed, if the recurrent 'madness of vision' that has been associated with the baroque is now enjoying an unprecedented wave of enthusiasm, as

Martin Jay proposes (1992: 189), then arguably its displacement of classical or Cartesian perspectives is on its way to becoming as integral to the discourses of technoscience as it is to those of aesthetics – if we can still usefully make such a distinction.

In order to come to terms with this sensibility in its latest manifestation, we need to look somewhere between the anciently interred traces of microbial promiscuity and the all-too-recent flourishing of electronic miscegenation. It is in the city – at the hubs of human movement and habitation – that we find a long but still relatively accessible history of socially accelerated flows and fusions, here that we might uncover a succession of culturally mediated human encounters with the aliens within and without. Before the Internet could be constituted as a luxuriating ecology of life-like entities, I would suggest, it was first necessary to the construe the city as a mesh of heterogeneous elements, to experience the variegated life secreted in *les passages* and *le paysage des grandes villes*; if not literally, then at least metaphorically. To a far greater degree than during its recent enmeshing with new electronic media, the human body in the metropolis has been open to diverse flows, has entertained new forms, has participated in a 'baroque sociability' with all its invited and uninvited guests.

To rethink these 'transhuman' interchanges in terms of new theories of open systems and non-linear dynamics is not to cede great tracts of social or cultural life wholesale to the methodologies and conceptual frameworks of the physical sciences. It is too well-recognized that the 'hard' sciences are themselves socio-culturally conditioned, too widely accepted that they will sooner or later be subject to further transformation ever to attempt such a brazen manoeuvre. But, by the same token, proposing that current scientific concerns with open systems and generative matter might be influenced by the experience of cosmopolitan life – or any other aspect of sociation – is not to draw these fields into the realm of sociological or culturalist determination: a gesture which the gravity of environmental concerns has helped relegate to a bygone era. It is, rather, to encourage some degree of contamination between the still formidable meta-disciplinary structures; to push for a porosity between the study of the social and the study of the natural that corresponds to the openness of the human organism to the greater flux of energy, matter and life.

Notes

1. For an illuminating discussion of the 'transhuman' – which tracks the concept from the work of Nietzsche and develops it in the light of the philosophy of Deleuze and Guattari, see Ansell Pearson (1997).
2. Although the evidence drawn on in this article is overwhelmingly European, there are interesting parallels to be found in the work of the Japanese architect and urban designer, Kisho Kurokawa. Kurokawa (1991) draws on such

scientific notions as hybridity, symbiosis and the rhizome in his celebration of the dense, multivalent nature of traditional Japanese cities. He also makes connections between the European baroque and the 'other modernity' of Japan.

References

Adam, Barbara (1993) 'Time and Environmental Crisis: An Exploration with Special Reference to Pollution', *Innovation in Social Science Research* 6(4): 399–413.

Ansell Pearson, Keith (1997) *Viroid Life: Perspectives on Nietzsche and the Transhuman Condition.* London: Routledge.

Aragon, Louis (1970) *Nightwalker (Le Paysan de Paris).* Englewood Cliffs, NJ: Prentice Hall.

Arnold, David (1996) *The Problem of Nature: Environment, Culture and European Expansion.* Oxford: Blackwell.

Bataille, Georges (1991) *The Accursed Share: vol. 1, General Economy.* New York: Zone Books.

Baudelaire, Charles (1964) *The Painter of Modern Life and Other Essays.* London: Phaidon.

Bauman, Zygmunt (1992) 'The Solution as Problem', *Times Higher Education Supplement* 1045 (13 November): 25.

Beck, Ulrich (1992) *Risk Society: Towards a New Modernity.* London: Sage.

Beck, Ulrich (1995) *Ecological Politics in an Age of Risk.* Cambridge: Polity Press.

Beck, Ulrich (1996) 'World Risk Society as Cosmopolitan Society? Ecological Questions in a Framework of Manufactured Uncertainties', *Theory, Culture & Society* 13(4): 1–32.

Benjamin, Walter (1973) *Charles Baudelaire: A Lyric Poet in the Era of High Capitalism.* London: New Left Books.

Benjamin, Walter (1977) *The Origin of German Tragic Drama.* London: New Left Books.

Benjamin, Walter (1978) *Reflections: Essays, Aphorisms, Autobiographical Writings.* New York: Harcourt Brace Jovanovich.

Brand, Stewart (1991) 'Foreword', in K.E. Drexler, C. Peterson and G. Pergamit (eds) *Unbounding the Future: The Nanotechnology Revolution.* New York: William Morrow.

Buci-Glucksmann, Christine (1994) *Baroque Reason: The Aesthetics of Modernity.* London: Sage.

Buck-Morss, Susan (1989) *The Dialectics of Seeing: Walter Benjamin and the Arcades Project.* Cambridge, MA: MIT Press.

Cilliers, Paul (1998) *Complexity and Postmodernism: Understanding Complex Systems.* London: Routledge.

Clark, Nigel (1998) 'Nanoplanet: Molecular Engineering in the Time of Ecological Crisis', *Time & Society* 7(2): 353–68.

Davis, Mike (1998) *Ecology of Fear: Los Angeles and the Imagination of Disaster.* New York: Metropolitan Books.

De Landa, Manuel (1992) 'Nonorganic Life', in J. Crary and S. Kwinter (eds) *Incorporations.* New York: Zone Books.

De Landa, Manuel (1993) 'Virtual Environments and the Emergence of Synthetic Reason', in M. Dery (ed.) *Flame Wars: The Discourse of Cyberculture*, Special Issue of *South Atlantic Quarterly* 92(4): 793–815.

De Landa, Manuel (1996) 'Markets and Antimarkets in the World Economy', in S. Aronowitz, B. Martinsons and M. Menser (eds) *Technoscience and Cyberculture.* New York and London: Routledge.

De Landa, Manuel (1997) *A Thousand Years of Nonlinear History.* New York: Swerve Editions.

Deleuze, Gilles (1993) *The Fold: Leibniz and the Baroque.* Minneapolis and London: University of Minnesota Press.

Deleuze, Gilles (1995) *Negotiations.* New York: Columbia University Press.

Deleuze, Gilles and Félix Guattari (1987) *A Thousand Plateaus: Capitalism and Schizophrenia*. Minneapolis: University of Minnesota Press.
Dibbell, Julian (1995) 'Viruses are Good for You', *Wired* 3 (February): 126–33, 172–80.
Eder, Klaus (1996) *The Social Construction of Nature*. London: Sage.
Featherstone, Mike and Roger Burrows (1995) 'Cultures of Technological Embodiment: An Introduction', in M. Featherstone and R. Burrows (eds) *Cyberspace/Cyberbodies/Cyberpunk: Cultures of Technological Embodiment.* London: Sage.
Ferguson, Priscilla Parkhurst (1994) 'The *Flâneur* On and Off the Streets of Paris', in K. Tester (ed.) *The Flâneur*. London and New York: Routledge.
Foucault, Michel (1991) *Discipline and Punish: The Birth of the Prison*. London: Penguin.
Frisby, David (1985) *Fragments of Modernity*. Cambridge: Polity Press.
Giddens, Anthony (1994) 'Living in a Post-Traditional Society', in U. Beck, A. Giddens and S. Lash (eds) *Reflexive Modernization: Politics, Tradition and Aesthetics in the Modern Social Order.* Cambridge: Polity Press.
Hannerz, Ulf (1980) *Exploring the City: Inquiries Toward an Urban Anthropology*. New York: Columbia University Press.
Haraway, Donna (1991) *Simians, Cyborgs and Women: The Reinvention of Nature*. New York and London: Routledge.
Horgan, John (1998) *The End of Science*. London: Abacus.
Jay, Martin (1992) 'Scopic Regimes of Modernity', in S. Lash and J. Friedman (eds) *Modernity and Identity*. Oxford: Blackwell.
Kauffmann, Stuart (1995) *At Home in the Universe: The Search for Laws of Complexity.* London: Penguin.
Kelly, Kevin (1994) *Out of Control: The New Biology of Machines*. London: Fourth Estate.
Kirby, Vicki (1997) *Telling Flesh: The Substance of the Corporeal.* New York and London: Routledge.
Kurokawa, Kisho (1991) *Intercultural Architecture: The Philosophy of Symbiosis.* London: Academy Editions.
Lash, Scott (1993) 'Reflexive Modernization: The Aesthetic Dimension', *Theory, Culture & Society* 10(1): 1–23.
Lash, Scott (1994) 'Reflexivity and its Doubles: Structure, Aesthetics, Community', in U. Beck, A. Giddens and S. Lash (eds) *Reflexive Modernization: Politics, Tradition and Aesthetics in the Modern Social Order.* Cambridge: Polity Press.
Leakey, F.W. (1969) *Baudelaire and Nature.* Manchester: Manchester University Press.
Lechte, John (1995) '(Not) Belonging in Postmodern Space', in S. Watson and K. Gibson (eds) *Postmodern Cities and Spaces.* Oxford: Blackwell.
Lévi-Strauss, Claude (1975) *Tristes Tropiques*. New York: Atheneum.
Luke, Timothy (1997) 'Digital Beings and Virtual Times: The Politics of Cybersubjectivity in Nicholas Negroponte's *Being Digital*', *Theory & Event* 1(1) [http://muse.jhu.edu/journals/theory_&_event/v001.1r_luke.html].
McKibben, Bill (1990) *The End of Nature*. London: Penguin.
Macnaghten, Phil and John Urry (1998) *Contested Natures*. London: Sage.
Maffesoli, Michel (1996) *The Time of the Tribes: The Decline of Individualism in Mass Society*. London: Sage.
Margulis, Lynn and Dorian Sagan (1995) *What is Life?* London: Weidenfeld and Nicolson.
Morin, Edgar (1974) 'Complexity', *International Social Science Journal* 26(4): 555–82.
Musil, Robert (1953) *The Man Without Qualities, vol. 1.* London: Secker and Warburg.
Plant, Sadie (1997) *Zeros + Ones: Digital Women + the New Technoculture*. London: Fourth Estate.
Prigogine, Ilya and Isabelle Stengers (1984) *Order out of Chaos*. New York: Bantam.
Quammen, David (1998) 'Planet of Weeds', *Harper's Magazine* 297(1781) 57–69.

Sagan, Dorian (1992) 'Metametazoa: Biology and Multiplicity', in J. Crary and S. Kwinter (eds) *Incorporations*. New York: Zone.
Shields, Rob (1994) 'Fancy Footwork: Walter Benjamin's Notes on Flânerie', in K. Tester (ed.) *The Flâneur*. London and New York: Routledge.
Smith, Neil (1984) *Uneven Development: Nature, Capital and the Production of Space*. Oxford: Basil Blackwell.
Spirn, Anne Whiston (1984) *The Granite Garden: Urban Nature and Human Design*. New York: Basic Books.
Stafford, Barbara Maria (1996) *Good Looking: Essays on the Virtue of Images*. Cambridge, MA: MIT Press.
Tester, Keith (ed.) (1994) *The Flâneur*. London and New York: Routledge.
Urry, John (1995) *Consuming Places*. London: Routledge.
Van Loon, Joost (1997) 'The End of Antibiotics: Notes toward an Investigation', *Space and Culture* 1(2): 127–48.
Wilson, Elizabeth (1995) 'The Invisible Flâneur', in S. Watson and K. Gibson (eds) *Postmodern Cities and Spaces*. Oxford: Blackwell.

Nigel Clark teaches Geography at the Open University. His last article in *Body & Society* was on cyberbodies (1995). He is working on a book on biophysical creativity and environmentalism, provisionally titled *Wild Life*.

Still Life in Nearly Present Time: The Object of Nature

NIGEL THRIFT

> If winds, currents, glaciers, volcanoes, etc., carry subtle messages that are so difficult to read that it takes us absolutely ages trying to decipher them, wouldn't it be appropriate to call them intelligent? How would it be if it turned out that we were only the slowest and least intelligent beings in the world? (Serres, 1995: 30)

> It is not enough to say the subject is constituted in a symbolic system. . . . It is [also] constructed in real practices. (Foucault, 1984: 369)

> . . . what happens if the half-second delay is set, not in a super-sensible domain, but in the corporealisation of culture and the culturalisation of corporeality? (Connolly, 1999: 20)

Introduction: Towards a Genealogy of Background

In this article, I want to make an argument concerning the importance of nature, the body and time in Western societies. It is not, I think, the usual kind of argument, based upon genealogical accounts of the rise and fall of discourses like romanticism or modernism (which is not to say that elements of these accounts do not adhere). Rather, it is an attempt to strike out towards a new understanding of how nature is apprehended, based upon giving much greater credence to that small but vitally significant period of time in which the body makes the world intelligible by setting up a background of expectation which, I will go on to argue, is much of what we feel as 'nature'.

In a sense, what I want to do is to restate some of the current concerns of the turn to a vitalist conception of the world. But I want to do so in a way which goes beyond the general and sometimes rather portentous philosophical statements

Body & Society © 2000 SAGE Publications (London, Thousand Oaks and New Delhi),
Vol. 6(3–4): 34–57
[1357–034X(200009/12)6:3–4;34–57;015469]

about time, the body and becoming which have now become so familiar (e.g. Grosz, 1999) by connecting up with understandings of body practices from the social sciences – and capitalist business.

Using these resources, I want to argue that nature has become a, and perhaps even the, key site of contemplation and mysticism in the modern world as a result of the evolution of a set of body practices which, as they have taken hold, have produced an expanded awareness of present time. My problem in making such an argument is that contemplation and mysticism are not practices much associated with an enhanced grasp of the modern world; they are more usually associated with figures from times of yore like hermits and monks. How can such practices of slowness make sense in an increasingly frantic capitalist world, a ferocious jumble of signals, journeys and screens which has squeezed out or is likely to squeeze out such sedate activity once and for all (see just most recently, Bertman, 1998; Brand, 1999; Flaherty, 1998; Kovach and Rosenstiel, 1999; Speak, 1999)? Surely it is all quick, quick and no slow.

In order to refute such easy characterization, I will therefore make an argument in six stages, each of which corresponds to a particular part of the article. The first part of the article therefore begins by setting out some theoretical aspirations, aspirations which all attempt to escape the traps of representational thinking of the kind that wants, for example, to understand nature as simply a project of cultural inscription (as in many writings on 'landscape') in favour of the kind of thinking that understands nature as a complex virtuality (Cache, 1995; Rajchman, 1998). With an account distilled from these thoughts, in the second part of the article I will argue that a go-faster world, in which time takes on an increasingly frenetic future-oriented quality, has been balanced by a series of contemplative practices – many of them to do with a heightened awareness of movement – which have, in fact, produced an expansion of awareness of the present. The third part of the article follows on. It concerns the classical idea that the world has been disenchanted. My argument here is to the contrary. In fact, the mystical qualities of the world remain in place. Assured by a whole series of body practices, some old and some new, these practices have produced an expansion of awareness of present time. The fourth part of the article then argues that the experience of these two sets of *immersive* body practices accounts for a large part of what we attend to as 'nature'; they define much of what we cleave to as a 'natural' experience by setting up a *background of expectation*. The fifth part of the article suggests that these body practices can be seen as part of a larger biopolitical project which is an attempt to renovate and value 'bare life'. But 'bare life' is not bare. It is most of what there is. Then, the sixth part of the article offers some words of warning. Another such project of renovation of bare life is already in motion, but it is being undertaken by business and its goal is a narrow one. The article concludes with some further clarifications.

Becoming There

My thinking on nature, the body and time in this article is based upon four different but quite clearly associated sources of inspiration which, when taken together, make it possible to construct an emergent account of emergent body practice which is the base of the rest of the article. The first of these is the work of biological philosophers and philosophical biologists like Clark (1997), Deleuze (1988), Margulis (1998), Margulis et al. (1999) and Ansell Pearson (1997, 1999) who want to argue for a reconfigured ethology in which bodies become means of transporting 'instincts'[1] which are best thought of as particular territories of becoming, maps mapping out 'populations' of identities and forces, zones and gradients, through differentiation, divergence and creation.

> Behaviour can no longer be localised in individuals conceived as preformed homunculi; but has to be treated epigenetically as a function of complex material systems which cut across individuals (assemblages) and which transverse phyletic lineages and organismic boundaries (rhizomes). This requires the articulation of a distributed conception of agency. The challenge is to show that nature consists of a field of multiplicities, assemblages of heterogeneous components (human, animal, viral, molecular, etc.) in which 'creative evolution' can be shown to involve blocks of becoming. (Ansell Pearson, 1999: 171)

Maps are their own practitioners, in other words.

The second source of inspiration is the revival of interest in the non-cognitive dimensions of embodiment. Probably 95 percent of embodied thought is non-cognitive, yet probably 95 percent of academic thought has concentrated on the cognitive dimension of the conscious 'I'. Without in any way diminishing the importance of cognitive thought (though certainly questioning its exact nature), we can conceive of non-cognitive thought as a set of embodied dispositions ('instincts' if you like) which have been biologically wired in or culturally sedimented (the exact difference between the two being a fascinating question in itself), action-oriented 'representations' which simultaneously describe aspects of the world and prescribe possible actions. There has, of course, been a considerable amount of work on body practices stemming from the work of authors as different as Mauss, Benjamin, Wittgenstein, Merleau-Ponty and Bourdieu (e.g. Taussig, 1994), which recognizes that much of human life is lived in a non-cognitive world. But I think it is fair to say that its implications are only now being worked through, most especially in areas like performance studies, feminist theory and non-representational geographies. In particular, when we say that human beings act to think or that they learn by doing, we need to refigure what we count as thought and knowledge. In particular, much cognitive thought and knowledge may, indeed, be only a kind of post-hoc rumination; 'to be aware of an experience means that it has passed' (Norretranders, 1998: 128). For example, most of the time, an action is in motion before we decide to perform it; our average 'readiness potential' is about 0.8 seconds, although cases of up to 1.5 seconds have been recorded.[2]

As McCrone (1999: 135) makes clear, none of this means that conscious awareness is just along for the ride. Rather, we can say that the non-conscious comes to be more highly valued – the 'not properly conscious impulses, inklings, automotisms and reflexive action' can no longer be regarded as trivial. And, at the same time, conscious awareness is repositioned as a means of scrutinizing and focusing these actions. To put it another way, what has been found is that the body has a number of ways of conjuring with time that work through structures of *anticipation*, the something to be known which is very often the result of the body's own movements, which leave 'some aspect of the movement standing proud' (McCrone, 1999: 158). Why? Because:

> The brain was never really designed for contemplating images. Our ability to imagine and fantasise is something that has to piggyback on a processing hierarchy designed first and foremost for the business of perception. And to do perception well, the brain needs a machinery that comes up with a fresh wave of prediction at least a couple of times a second, or about as fast as we can make a substantial shift in our conscious point of view . . . it would be unnatural for the brain to linger and not move on.

In turn, such work points to the pivotal importance of emotions as the key means the body has of sorting the non-cognitive realm through a range of different sensory registers, including the interoceptive (including not only the viscera but also the skin), the proprioceptive (based on musculo-skeletal investments) and fine touch which involves the conduct of the whole body and not just the brain.

> Note that, depending on the object, there may be different proportions of musculoskeletal and emotional accompaniment, but both are always present. The presence of all these signals . . . describes both the object as it looms towards the organism and part of the reaction of the organism towards the object . . . (Demasio, 1999: 147)

Which brings us to the third source of inspiration – the much greater emphasis that is being placed on the object. To begin with, the body is objectified as a composite of biological-cultural 'instincts' which enable and in many ways constitute thought as a result of the development of particular organs. For example, the development of the hand, with all the possibilities it presents, was an impetus to the redesign, or re-allocation of the brain's circuitry so that the hand speaks to the brain just as much as the brain speaks to the hand (Wilson, 1998).

Then, these organs are closely linked with particular objects (Sudnow, 1993). Organs like the hand become as one with tools they relate to.

> The idea of 'becoming one' with a [mechanical] back hoe is no more exotic than the idea of a rider becoming one with a horse or a carpenter becoming one with a hammer, and this phenomenon may itself take its origin from countless monkeys who spent countless eons becoming one with tree branches. The mystical feel comes from the combination of a good mechanical marriage and something in the nervous system that can make an object external to the body feel as if it had sprouted from the hand, foot, or (rarely) some other place on the body where your skin makes contact with it. (Wilson, 1998: 63)

Then again, objects do not just constitute an extension of bodily capacities; they themselves are a vital element in distributed ecology of thought, so that 'what used to look like internalisation (of thought and subjectivity) now appears as a gradual propagation of organised functional properties across a set of malleable media' (Hutchins, 1995: 312); 'the true engine of reason . . . is bounded neither by skin nor skull' (Clark, 1997: 69). Thus, as Hayles (1999: 290) puts it:

> . . . no longer is human will seen as the source from which emanates the mystery necessary to dominate and control the environment. Rather, the distributed cognition of the emergent human subject correlates with – in Bateson's phase, becomes a metaphor for – the distributed cognitive system as a whole, in which thinking is done by both human and nonhuman actors.

Not only do objects make thought do-able (e.g. Latour and Hermant, 1998) but they also very often make thought possible. In a sense, then, as parts of networks of effectivity, objects think. We might even go still further, by arguing that 'everything that is resounds' (Lingis, 1998: 99):

> It is not that things barely show themselves, behind illusory appearances fabricated by our subjectivity; it is that things are exorbitantly exhibitionist. The landscape resounds; facades, caricatures, halos, shadows, dance across it. Under the sunlight extends the pageantry of things. The twilight does not put an end to their histrionics. In the heart of the night the pulse of the night summons still their ghosts. (Lingis, 1998: 100)

The fourth source of inspiration is the genealogy of the body practices which must be a large part of an ethology of 'instincts' – these are now, finally, coming under intense scrutiny. Grouped around terms like 'performance', and around theorists like Bourdieu and Foucault, researchers in the social sciences and humanities have, over the last 20 years, begun to produce a history of particular organs (e.g. Hillman and Massio, 1997; Jordanova, 1994) and particular body practices – from drill to dance. But it is true to say that we still understand very little of how the body practices that comprise 'us' have come down to and inhabit us, passing into our being, passing our being back and forth between bodies and passing our being on (Hayles, 1999).[3]

These four sources of inspiration allow us to begin to sense, through this combination of work in areas as diverse as biology, philosophy and performance studies, what an understanding of that little space of time that is much of what we are, a space not so much at the edge of action as lighting the world. I will call this domain 'bare life' after Aristotle's notion of *Zoé*, a 'simple natural sweetness' (Agamben, 1998). Of course, it is not really bare; bare life pulses with action. And it is not simple. And it is not preternatural. But what such a notion allows us to do is to point beyond the grand notions of bodily hexis like *habitus* towards something more specific and more open to description. And it does two more things. One is to begin to understand qualities like anticipation and intuition as not just spirits but material orientations. And the other is to understand that this little space of time is a vast biopolitical domain, that blink between action and performance in which the world is pre-set by biological and cultural instincts which bear both extraordinary genealogical freight – *and* a potential for potentiality.

How might we begin to understand the structure of this domain of flourishing? One manageable and useable account has been offered by Gil (1998). Gil argues, as I would (see Thrift, 1996, 1997, 1998, 2000), that we need to escape the constructionist notion of the body as simply an inscribed surface, in which the body is reduced to what Gil (1998) calls a 'body image', an individual unitary, organismic body which can act as a surface upon which society can construct itself. This interpretation is mistaken in at least three ways. First, the body becomes a static signified to be filled with signs of society. Second, the body is divorced from other things, from the object world. Third, the body is located *in* space, it does not produce space. But, there is another, non-representational view (Thrift, 1997, 2000). In this view, the body is 'not about signs and meanings but about a mechanics of space' (Gil, 1998: 126) brought about by the relation between bodies and things. Thus:

> ... the space of the body has limits that are not those of the body image, if we understand by that the limits of the body lived in a unitary fashion. The limits of the space of the body are in things. In movement, for example, the body places changing limits on these things. To the extent that they are 'subjective', these limits constitute the end result of the integration into the body of the relations (of distance, form, and so on) that it holds with things in objective space. To the extent that they can be pinned down topologically, these limits are no longer 'lived' but are properties of space itself. (Gil, 1998: 125)

In turn, and following a Deleuzian interpretation:

> ... the body 'lives' in space, but not like a sphere with a closed continuous surface. On the contrary, its movements, limbs and organs determine that it has regular relations with things in space, relations that are individually integrated for the decoder. These relations imply *exfoliations* of the space of the body that can be treated separately. Relations to a tree, a prey, a star, an enemy, a loved object or desired nourishment set into motion certain privileged organs including precise spaces of the body. Exfoliation is the essential way the body 'turns on to' things, onto objective

> space, onto living things. Here there is a type of communication that is always present, but only makes itself really visible in pathological or marginal experiences. Nevertheless the ordinary experience of relations to things also implies this mode of communication. Being in space means to establish diverse relationships with the things that surround our bodies. Each set of relations is determined by the action of the body that accompanies an investment of desire in a particular being or particular object. Between the body (and the organs in use) and the things is established a connection that immediately affects the form and space of the body; between the one and the other a privileged spatial relation emerges that defines the space uniting them as 'near' or 'far', resistant, thick, wavy, vertiginous, smooth, prickly. (Gil, 1998: 127)[4]

In other words, the space of the body consists of a series of 'leaves', each of which 'contains' the relations of the body to things and each of which is more or less related to other spaces. Correspondences are not, at least initially, conceptual but result 'from the work done by the body spatialising space' (Gil, 1998: 130). Thus:

> Analogy, similitude, opposition, and dissimilitude are given in the forms of the space of the body before being thought of as concepts. In the same ways as the 'concrete science' which establishes classification on the basis of sensorial differences found in 'primitive thought', the recording-body gathers up, brings together, unites, dislocates, spreads, and separates things to the spatial forms that contain in themselves (because they bring them about) the properties of unification and division. (Gil, 1998: 130)

It follows that in what follows body practices are *not* to be thought of, at least in the first instance, as cognitive. This would be a first-order mistake. For, to reiterate, we know that 'consciousness is a measure of but a very small part of what our senses perceive' (Norretranders, 1998: 127).

> Conscious thought is the tip of an enormous iceberg. It is the rule of thumb among cognitive scientists that unconscious thought is 95 per cent of all thought – and that may be a serious underestimate. Moreover the 95 per cent below the surface of conscious awareness shapes and structures all conscious thought. If the cognitive unconscious were not doing this shaping, there could be no conscious thought.
>
> The cognitive unconscious is vast and intricately structured. It includes not only all our automatic cognitive operations, but also all our implicit knowledge. All of our knowledge and beliefs are formed in terms of a conceptual system that resides mostly in the cognitive unconscious.
>
> Our unconscious cognitive system functions like 'a hidden hand' that shapes how we conceptualise all aspects of our experience. This hidden hand gives form to the metamorphosis that is built into our ordinary conceptual system. It creates the entities that inhabit the cognitive unconscious – abstract entities like friendships, bargains, failures and lies – that we use in ordinary unconscious reasoning. It thus shapes how we automatically and unconsciously comprehend what we experience. It constitutes our unreflective common sense. (Lakoff and Johnson, 1998: 13)

And this cognitive unconscious rises out of the layerings and interleavings of body practices and things which we might frame as 'instincts' or, more accurately, as structured *anticipations*. For neither is conscious awareness just going for the ride on the back of the cognitive unconscious. Rather, every moment is processed as a prior intent, style or tone which arises from perception-in-movement, every moment is the fleeting edge of a sensory forecast (McCrone, 1999), quite literally a stance to the world.

With these 'thoughts in mind' (how easily we use these questionable phrases), we can now move to a consideration of how body practices show up in the modern world and how the modern world shows up in them. To do this, we first need to clear away some tired old pictures of the world.

The Go-Faster World

Elsewhere (Thrift, 1995, 1996, 1997), I have criticized the notion that we live in a speeded-up world in which friction has been lost and everyday life skids along on the plane of velocity. Much of the literature which enforces this notion is based upon a simple technological determinism which unproblematically maps the apparent powers of things on to subjects. While it is undeniable that people and messages now move faster than they did, old practices have been adjusted, and new practices have been invented, which make it impossible to simply read off this physical fact on to culture. Further, it is possible to argue that speed is itself in part a cultural creation, a classical modernist trope now in general cultural circulation (see Kern, 1983) as a series of metaphors and analogies and as a rhetoric of 'speedy' things.[5] This cultural creation of speed itself depends upon the depiction of certain places, things and people as slow-moving, most particularly those places, things and people connected with nature, the countryside and so on.

This is, of course, a very strange opposition since one might just as well argue – precisely through the instruments which have become available to measure speed – that nature is actually very fast. The speed of light is, well, the speed of light. Chemical reactions can work at astounding speeds. Even that slow old thing, the human body, works reasonably fast. Though in our brains, nerve impulses only tend to crawl along – at between 2 and 20 miles per hour – along the heavily myelinated nerves (such as muscle and the sensory nerves) nerve impulses travel at up to 240 miles per hour (McCrone, 1999; Norretranders, 1998).

But, more than this, the opposition ignores a general reconstruction of time which has taken place (quite literally) over the last 150 years, a sense of body practices which constitute and value *the present moment*, rather than spearing into the future. Ironically, these body practices have all taken shape around the increasing awareness of kinaesthesia, a sixth sense based on the interactive *movement* and subsequent awareness of body parts:

> . . . we obtain a sense of our own movement not only from specialised receptors in the inner ear, joints, tendons and muscles but also from what we can see, hear and feel . . . vision, 'for instance', is kinaesthetic in that it registers movements of the body, just as much as do the vestibular receptors and those in the muscles, joints and skin. (Reason, 1982: 233)

Contrary to Sherrington's direct correlation of sensory experience with the activation of specific receptors and their nerves of different cellular levels, the kinaesthetic sense is a gestalt emerging from the interaction of all the other senses. After Gibson we can speak of kinaesthesia in terms of its muscular, articular, vestibular, cutaneous, auditory and visual modalities (Gibson, 1966: 36–8). In this view kinaesthesia is the ground to our consciousness (Stewart, 1998: 44).

I think it can be argued that greater awareness of movement has in turn produced a set of resources that enable us to separate out and value a present-orientated stillness, thus promoting a 'politics' based in intensified attention to the present and unqualified affectivity. Where might this present-orientation have come from? I would argue that its history is born out of a number of developments which, taken together, constitute a genealogy of the present.

The first of these developments is practices of contemplation. Foucault and others have highlighted the significance of confession as a model for recent practices of the technology of the self. I think an argument can be made for a similar kind of history based in practices of contemplation understood as 'aptitudes of performance' (Asad, 1993), rather than explicit belief. This history might touch upon certain forms of prayer, the practices of some rituals and other religious technologies which concentrate time.

Whatever the case, there seems no doubt that extant practices of contemplation were gradually transmuted by a whole series of developments in the 19th century and thereafter (Segel, 1998). The first of these was the development of a series of body practices which stressed sensory appreciation through a more complete control of the body in order to provide more harmonious relations with the environment. A good example of these developments is the rise of various body techniques like the Alexander Technique, the Feldenkrais technique, Bioenegetics and Body–Mind Centring, which teach movement awareness and the reorganization of movement sensation (Feldenkrais, 1972; Hartley, 1995; Lowen, 1975; MacDonald, 1998; McGowan, 1997a, 1997b). Feldenkrais (1972), for example, argued that cultivation of certain bodily practices could enhance our ability to 'know' the world through systematic correction of what he called the 'body image'.

The second development is the rise of systematic knowledge of body measurement, based on increasing the efficiency of the body. From Marey's and

Muybridge's study of the physiology of movement through to Gilbreth and Taylor's time-and-motion studies and modern ergonomics and sports science, the study and articulation of minute human movement have become a key to producing human comportment (Dagognet, 1992; Mattelart, 1996). In turn, it can be argued that the increasingly fine grain of the many bodily movements built out of this study has made its mark on how time is constructed by the body.[6]

The third development is the fixing of a still, contemplative gaze, which is able to capture transience. Such a gaze is found in art from the 18th century on but reaches a kind of technological fulfilment in the photograph, especially with the growth of popular photography from the end of the 19th century onwards. Crawshaw and Urry (1997) argue that popular photography consists of a set of socially organized rituals which fix a place, a 'language' of material objects through which we understand and appreciate the environment (and the material objects themselves) and a means of organizing time itself. In each case, what are being described are a set of practices which momentarily fix the body and other things in spaces and times by producing spaces and times in which they can be fixed.[7]

The fourth development is the forging of a body of knowledge about social interaction as the distillation of detailed body practices. Such knowledge can already be found in the 19th century and early 20th century (for example, in the development of various movement notations) but it reaches a peak in the 20th century with the rise of various knowledges of body practice from work on the psychology of body language and gesture, though work on bodily intonations of space, as in Hall's 'proxemics' (Hall, 1990), through to the detailed conversational analysis of symbolic interactionism, ethno-methodology and the like to be found in the work of Goffman, Garfinkel, Sacks and so on (e.g. Burns, 1992). In turn, this knowledge, much of which was developed in academia and other relatively formal arenas, has gradually seeped out into everyday life as a whole new corporeal curriculum of expressive competence, for example through courses on body language (now being given, for example, to checkout operators in some supermarket chains), cultural awareness training and all manner of training in self-presentation (cf. Giddens, 1991; Thrift, 1997). Thus, what was quite specific knowledge has become general and routine.

Each and every one of these four developments of body practice stretches out the moment, most especially by paying detailed attention to it. They expand, if you like, the 'size' of consciousness, allowing each moment to be more carefully attended to and invested with more of its context. Taken together, they may be seen as constructing a slow-down of perception, as much as a speed-up.

Re-enchantment

These developments have to be taken in concert with others to complete my argument. One of the most damaging ideas that has swept the social sciences and humanities has been the idea of a disenchanting modernity (Thrift, 1996). This act of purification has radically depopulated thinking about Western societies as whole sets of delegates and intermediaries have been consigned to oblivion as extinct impulses, those delegates and intermediaries which might appear to be associated with forces of magic, the sacred, ritual, affect, trance and so on. Now, however, the contemporary turn towards vitalist ways of thinking (cf. Watson, 1998b) has made it much easier to see that the magic has not gone away. Western societies, like all others, are full of these forces (Dening, 1996; Muecke, 1999). They can be seen as concentrating, in particular, in a set of practices which can be described as 'mystical'. Like practices of contemplation, with which they are intimately linked, they can be seen as the result of a number of overlapping processes of animation and play which allow forces and intensities to be focused and channelled: it is *stimulation* that produces tranquillity and it is *stimulation* that produces trance.

First, then, there is the importance of various forms of mystical communication, mental and physical techniques that 'fix the conditions of possibility of an encounter or dialogue with the other (method of prayer, meditation, concentration)' (de Certeau, 1992: 5). Current forms of practice have a long genealogy in Western cultures and stem from traditions as different as the Christian (both Anglican and Catholic), the nature mysticisms of Romanticism as found in various forms of the sublime, the numerous forms of Eastern thinking which have been imported into the West, especially in the 19th century, and the cathartic elements of many types of performance. More recently, there has been the growth of New Age religions, nearly all of which contain an explicitly mystical component (for example, following on from sources as diverse as the writings of Gurdjieff or Hopi Indian practices). Not least, in all these traditions can be found, to a greater or lesser degree, an approach to nature as both the focus and the object of mystical energies. For example, New Age thinking often stresses grids of power like ley lines, nature goddesses and the like, as well as the importance of particular sites as magical territories able to conjure up communication with the other.

This brings us to the second process, the importance of ritual, understood as practices which offer a heightened sense of involvement in our involvements through various performative technologies (Hughes-Freeland, 1998; Schechner, 1995). There may actually have been a multiplication of these performative spaces of affirmation, in which mystical experiences can be brought forth and animated through the power of body postures, repetitive movements, schedules of recall and spatial juxtapositions. Western societies have evolved more and more bodily practices which are a means of amplifying passions and producing 'oceanic' experiences: music, dance, theatre, mime, art and so on, which very large numbers of the population participate in; rather more than is often thought (see, for example, Finnegan, 1989). These practices have at least the potential to provide mystical experiences[8] – the trance state of some kinds of dance (Malbon, 1999), the 'high' of listening to a piece of music, and so on.[9]

Last, but not least, there has been the rise of varying forms of body therapy, which, though they often rest on various psychological and psychiatric principles, have quite clear links not only to contemplative but to mystical body practice. These are the various forms of dance therapy (e.g. Roth, 1998), music therapy, massage therapy, variants of bioenergetics (e.g. Lowen, 1975), autogenic therapy, body–mind centring, and so on, which try to harness and work with emotional energy on the grounds that movement causes emotion, rather than vice versa.

These body practices again allow the present to be intensified since they produce both an intensified sense of body movement and, at the same time, focus and enhance that movement. They are tempos of involvement without any necessary intention or initiative. They 'flow' time through the minute particulars of body movements that both have effects and yield experiences. They are 'performed dreams' (Schechner, 1995), 'virtual actualizations' of time which allow consciousness to become acute without necessarily being directed by drawing on the non-cognitive.

Nature as Background

What I want to argue next is that these contemplative and mystical developments which, taken as a whole, are widespread in modern Western societies, constitute a *background* within which nature is apprehended and which provides quite particular experiences of what nature is. They form, if you like, an embodied 'unconscious', a set of basic exfoliations of the body through which nature is constructed, planes of affect attuned to particular body parts (and senses) and corresponding elements of nature (from trees and grass, to river and sky) (Massumi, 1996), 'the sense and recognisability of things . . . do not lie in conceptual categories in which we mentally place them but in their positions and orientations which our postures address' (Lingis, 1998: 59).

Following on from this point, I want to argue, very tentatively, that these immersive practices are producing a new form of vitalism (Watson, 1998a, 1998b), a stance to *feeling* life (in the doubled sense of both a grasp of life, and emotional attunement to it) which explain many of the strong, sometimes even fanatical, investments that are placed on the 'natural'. The very ways in which, through these practices of contemplation and mysticism, embodiment is reproduced in the West, have produced an increasing bias towards framing life as a moving force, as push. In other words, the forms of embodiment I have set out in this article constitute a biopolitical domain arising out of a heightened awareness of particular forms of embodiment which, in turn, allow certain forms of signification to be grasped 'instinctively'.

This biopolitical domain has been strengthened by three developments. The first of these is the turning of certain body practices into privileged kinaesthetic spaces, and the privileged kinaesthetic spaces into body practices. I am thinking here especially of *walking* which since the 19th century, precisely in association with greater mobility (Wallace, 1993), has produced a new experience of nature. This is not walking as travel, but walking for itself. As walking becomes a natural practice to be indulged in for its own sake, so, against the background I have outlined, it can become a means to contact the Earth, to be at one with 'nature', even to be deemed therapeutic. It becomes a means of gathering stillness, without having to stay still, a means of contemplation and mystical communion to be found within the body. Lingis (1998: 70) captures what I take to be a culturally particular investment particularly well:

> . . . when we go out for a walk, our look is not continually *interested*, surveying the environment for landmarks and objectives. Even when we are on our way somewhere, for something, once launched we shift into just enjoying – or ending – the walk or the ride. Our gaze that prises beyond things is not situating on coordinates. It surveys across things, drawn to the distance when it fuses into the tone and mists of space. . . . The perception of things, the apprehension of their content and of their forms, is not an appropriation of them, but an expropriation of our forces into them, and ends in engagement.

Of course, none of this is to deny the cultural industry that has grown up around the practice of walking – the vast literature of books and guides, the special clothing and so on, all of which enhance or expand the range of affordances that inhere in any setting – but it is to suggest that the power of the meanings circulated by this industry is founded in the intensification of present experience coded in the body practices set out above. The background has allowed this foreground of symbolic delegates to develop.

The second development is, as the example of walking shows, the style of the body's location in space. What has developed has been both overall body stance and the formation of certain sequences of bodily experiences which, in their virtualized nature, produce an *expectation*, *an anticipation*, of a 'natural' experience: 'it is the way in which the body sits in space that allows signification to be grasped' (Gil, 1998: 109). For example, travel to a 'natural' place sets up the body to fall into a 'natural' stance to the world. There is, if you like, a genetics of movement which the body slips into through constant practice. There are 'dance floors of nature' (Lingis, 1998: 87).

The third development is similar but different. The body attends to configurations of objects which are in line with its expectations and which produce particular exfoliations/spaces and times. The body produces spaces and times through the things of nature which, in turn, inhabit the body through that production. Thus, for example, trees do not so much *mean* nature (Rival, 1998) as they are present as evidence of a natural configuration that embodiment itself has produced: our bodies know themselves in such thinking. Thus trees become flesh by being bound up in a practical field. And, in the intensified present time I have described, that presence becomes its own justification. There *is*. Nature, in a sense, becomes more natural.

In turn, of course, nature, understood as body practices like walking, expectations and configurations of objects, pushes back in confirmation. For example, our experience of walking is validated by its effects on the body – from sweat to heart rate to muscles stretching – which are a function of a *resistance* on certain planes which confirms the *existence* of other planes. So nature speaks in us as 'an infralanguage' (Gil, 1998) of movement which, through the articulations and micro-articulations of the body-in-encounter, fixes 'symbolic' thought as affect, mood, emotions and feelings[10] (thus as self-evidently present and numinous). Nature observes and writes us, bumping intensities into our thought[11] (understood especially as unconscious thought), rather as Deleuze would have it:

> [Deleuze's] projection of virtual elements too fast and multiple for conscious inspection or close third-person explanation meshes with his exploration of how differential degrees of intensity in thought moves it in some directions rather than others, open up lines of flight through which new concepts are introduced into being, and render thinking too layered and unpredictable to be captured by a juridical model in the Kantian tradition. He translates the story of juridical recognition in which Kant encloses thought in the last instance into one in which thinking is periodically nudged, frightened or terrorised into action by strange encounters. Recognition is a secondary formation often taken by consciousness in its innocence to be primary or apodictic, but thinking sometimes disturbs or modifies an established pattern of thought. (Connolly, 1999: 24)

'May I Not be Separated from Thee'[12]

In an important book, Giorgio Agamben (1998) manages to conjure up a depiction of 'bare life' (*zoé*) immured. Through the development of a whole set of governmental templates in a manner familiar to those who read Foucault or study the totalitarian state, bare life has become 'a life that has been deadened and mortified into juridical role' (Agamben, 1998: 187), a life 'naturalized' (to use a bitterly ironic term) from birth. Thus:

> . . . the Foucauldian thesis will have to be corrected or, at least, completed, in the sense that what characterizes modern politics is not so much the illusion of *zoé* in the polis – which is, in itself, absolutely ancient – not simply the fact that life as such becomes a principal object of the projections and calculations of state power. Instead the decisive fact is that, together with the process by which the exception everywhere becomes the rule, the realm of bare life – which is originally situated at the margins of the political order – gradually begins to coincide with the political realm, and exclusion and inclusion, outside and inside, *bios* and *zoé*, right and fact, enter into a zone of incredible indistinction. At once excluding bare life from and capturing it within the political order, the state of exception actually constituted, in its very separateness, the hidden foundation on which the entire political system rested. When its borders began to be blurred, the bare life that dwelt there frees itself in the city and becomes both subject and object of the political order, the one place for both the organization of state power and emancipation from it. Everything happens as if, along with the disciplinary process by which state power makes man as a living being into its own specific object, another process is set in motion that in large measure corresponds to the birth of modern democracy, in which man as a living being presents himself no longer as an *object* but as the *subject* of political power. These processes which in many ways oppose and (at least apparently) bitterly conflict with each other – nevertheless converge insofar as both concern the bare life of the citizen, the new biopolitical body of humanity. (Agamben, 1998: 8–9)

For Agamben, one of the questions is how to produce a notion of bare life that constitutes a politics but does not weigh it down with state imperatives. But his answer is pessimistic. Such a revitalization of bare life cannot be born.

> Bare life remains included in politics in the form of an exception, that is, as something that is included solely through an exclusion. How is it possible to 'politicize' the 'natural sweetness' of *zoé*? And first of all does *zoé* really need to be politicized, or is politics not already contained in *zoé* as its most priceless centre? The biopolitics of both modern totalitarianism and the society of mass hedonism and consumerism certainly constitute answers to these questions. Nevertheless, until a completely new politics – that is, a politics no longer founded on the *exception* of bare life – is at hand, every theory and every praxis will remain improvised and immobile and the 'beautiful day' of life will be given citizenship only either through blood and death or in the perfect senselessness to which the society of the spectacle condemns it. (Agamben, 1998: 11)

What Agamben seems to argue for, in part, is a revitalization of the body, in new forms of life: 'we are not only, in Foucault's words, animals whose life as living beings is at issue in their politics, but also – inversely – citizens whose very politics is at issue in their natural body' (Agamben, 1998: 188). What this article has argued is that such an emancipatory politics of bare life, founded in practices such as contemplation and mysticism, both already exists – and continues to come into existence in new ways – a politics founded especially but not only in a 'nature' which is a 'product of the double investment of the body by space (the information coming from the physical world) and the investment of space by the body (as a certain kind of receiver-encoder of information)' (Gil, 1998: 28). This is a politics of enhancement of the anticipation and conduct of certain bodily skills which, at the same time, contains its own premises though the effects of those skills.

This 'politics of the half-second delay' has the potential to expand the biopolitical domain, to make it more than just the site of investment by the state or investments by transnational capitalism. It may well explain the deep affective investments that are made by so many in a politics of nature, investments which move far beyond the cognitive and which are often figured as a restitution of all that has been lost. Perhaps, though, as this article has argued, the outcome might be figured more accurately as new appreciations and anticipations of spaces of embodiment, best understood as a form of magic dependent upon new musics of stillness and silence able to be discovered in a world of movement.

But: 'Step Inside the Great Outdoors'[13]

But, let's not overdo this. There are powerful contra-forces. For there is another politics of bare life which I have so far only touched on. This is the politics that arises out of the enormous efforts currently being made to foreground the background of bare life – to make it comprehensible and therefore able to be apprehended and so *made more of* – across a range of different interests and arenas. And of these interests and arenas the most powerful and, in many ways, the most advanced is capitalist business: Agamben's mass consumerism.

Capitalist firms are drawing on the various knowledges of bare life they are producing to produce new products, products which animate – 'turn on' – the body by producing an engaging and compelling ethology of the senses.

This is the rise of an 'experience economy' (Pine and Gilmore, 1999), a new genre of economic output which can construct experiences in order to produce added value. What have been the chief knowledges of bare life from which this experience economy has grown? There are four. The first has been tourism. Since the 1960s, a new kind of tourism has emerged based upon the theming of spaces. Relying on the experience of running museums, heritage centres, theme parks and certain kinds of themed retailing (Gottdiener, 1997) it has gradually constructed knowledge of how to produce spaces which can grip the senses. Of late, the kinaesthetic element of tourism has accordingly been amplified. For example, there are all the postcolonial forms of adventure set out by Guttman: houseboating, portaging, mountain-biking, cattle-driving, bob-sledding, tall-ship sailing, tornado-chasing, canyon orienteering, wagon training, seal viewing, iceberg tracking, racing car driving, hot-air ballooning, rock climbing, spelunking, white water rafting, canoeing, heli-hiking, hut-to-hut hiking, whale kissing, llama trekking, barnstorming, land yachting, historic battle re-enactments, iceboating, polar bearing and dog-sledding. The second knowledge, one clearly linked to the former, is sport and exercise. Sport and exercise have become key elements in modern experience economies, through their ability to influence bodily comportment (including specialized precision knowledges) through the specialized spaces that are constructed to serve them, and through the connections to the mass media (Abercrombie and Longhurst, 1998).

A third knowledge has been of performance. Since the 1960s again, knowledge of performance – which is, after all, extensive – has moved out from the stage to fill all manner of venues – from corporate presentations to the street. Buoyed up by mass media which have, in all probability, made the population at large more performative (Abercrombie and Longhurst, 1998), the art of performance has become a general art which concentrates especially on the conduct of the now and which can be appropriated. The fourth knowledge has been from education. Pedagogy has become a more and more active affair. Bolstered by findings from fields like cognition and consciousness, learning is now universally practised as active, even sensuous.

Capitalist firms have taken these knowledges and produced a series of purchases on the world. The first of these has been advertising. Advertising companies have become alive to an approach that takes in all the senses. Companies like the London advertising agency St Lukes have led the way towards advertising which is meant to tug at bare life by emphasizing kinaesthetic qualities. Another purchase is through sensorializing goods – producing goods that will richly engage the senses.

> Doing so requires awareness of which senses most affect customers, focuses on those senses and the sensations they experience, and the corporate redesign of the good to make it more appealing. Automakers, for example, now spend millions of dollars on every model to make sure that car doors sound *just so* when they close. Publishers greatly enhance the covers and interiors of books, and magazines with a number of tactile innovations (embossed lettering, scratching, bumpy or ultrasmooth surfaces) and sight sensations (translucent covers, funky fonts, clear photographs, three-dimensional graphics). Even presentation markers aren't just coloured anymore; Sanford scents them as well (liquorice for black, cherry for red, etc.). (Pine and Gilmore, 1999: 18)

Even quite simple goods are being designed which can feed back to the senses. For example, 'radar' baseballs make it possible to know how fast a ball was thrown, and generate social interaction since the catcher has to relay the speed back to the thrower.

The third purchase is the growth of packaged experiences which rely on theming contexts, so as to produce enhanced sensory experiences. This packaging can range all the way from the increasing outsourcing of children's parties from the home to companies, to the most elaborate virtual environments, which are virtually self-contained ethologies.

> Companies that want to stage compelling capacities should ... determine the theme of the experience as well as the impressions that will convey the theme to guests. Many times, experienced stagers develop a list of impressions they wish guests to store away and then think creatively about different themes and storylines that will bring the impressions together in the cohesive narrative. Then they winnow the impressions down to a manageable number – only and exactly those which truly devote the chosen theme. Next they focus on the animate and inanimate cues that could connote each impression, following the simple guidelines of accentuating the positive and eliminating the negative. They then must meticulously map out the effect each cue will have on the five senses – sight, sound, taste and smell – taking care not to overwhelm guests with too much sensory input. Finally, they add memorabilia to the total mix, extending the experience in the customers' mind over time. Of course, embracing these principles remains, for now, an art form. But those companies which figure out how to design experiences that are compelling, engaging, memorable – and rich – will be the ones leading the way into the emerging Experience Economy. (Pine and Gilmore, 1999: 61)

The fourth purchase is on objects that will produce kinaesthetic experiences, on the grounds that these experiences are usually the most compelling and the most memorable. What is fascinating is the speed with which this kinaesthetic purchase on the world is now expanding its grip, as knowledge of movement becomes engineered in institutions as different as film animation and special effects houses, virtual reality games, exponents of light shows, producers of extreme sports, and those who construct theme park rides. Increasingly, in particular, this knowledge is projected through objects which are based on maximizing movement experiences through the application of particular sequences of movement which engage the visceral sense as well as the proprioceptive and fine touch, rather like hieroglyphs of the kind found in dance and other performing arts (Thrift, 2000). For example, roller coasters are now often described in specifically choreographic terms.

Then a final purchase is, as already prefaced, memorabilia. Memorabilia both encapsulate and string out experiences. Most experience businesses mix memorabilia into what they offer. Memorabilia are becoming more sophisticated as objects can increasingly be customized. For example, guests' credit card signatures can be digitized and transferred to objects like clothing, sports equipment and photographs, often next to the signatures of appropriate celebrities. And, increasingly, memorabilia are being played for affective capacity. For example:

> Hillenbrand Industries of Butestaffe, Indiana, developed a new memorabilia capacity for the funeral industry. The concept emerged from the practice in many funeral homes of producing

> memory books for display at viewing and memorial services. Hillenbrand sought to bring greater efficiencies to the process but also to preserve the kind of one-of-a-kind collages families now put together to commemorate the lives of lost loved ones. Hillenbrand does this by developing a proprietary system to digitize, merge and print mass customized collages to both paper and video output media.
>
> But these life space collages serve merely as a prop for the experience Hillenbrand really offers. A self-guided kit that walks a family, group of friends or co-workers through a series of steps to create their own memories. 'What we sell', says Gary Bonnie, who handled the initiative, 'is the life scaping experience of gathering with others, rummaging through old photographs and other mementos, and recalling fond memories. The collage gift happens to be the outcome; the value is experienced in going through the process we've helped script.' Accordingly, Hillenbrand charges for the kit experience, whether or not people actually buy the collage. (Pine and Gilmore, 1999: 58)

So, what we see is bare life laid bare and anatomized, and put together again as saleable, immersive experiences. Through history, of course, landscapes have been constructed and experiences have been put up for sale but I think the new developments which, by engaging all the senses, produce new realms of experience to exchange should give us pause. It may be that 'the history of economic progress consists of charging a fee for what once was free' (Pine and Gilmore, 1999: 67). Alternatively, this maxim can be seen as simply another rationalization of the neoliberal order, one which entails a significant broadening and deepening of economic relations through much more sophisticated means of interpellation.[14]

In particular, of course, it involves a stance to nature, one which by *re-embodying* natural ethologies, using the examples gleaned from museums, theatre and theme parks, sets aside the immersive practices of contemplation and mysticism based on make-believe for immersion of a different kind based on make-us-believe (Walton, 1990). This is play without play, if you like[15] – play without the kind of anticipations that make live – that can produce an enhanced nature.

> In 1996 Ogden [Corporation] committed $100 million to create eight attractions called the American Wilderness Experience. There it immerses guests in nature scenes that feature the live animals, foliage, scents, and climates indigenous to various locales. The company's first American Wilderness Experience opened in late 1997 in the Ontario Malls Mill in San Bernadino, California. The company charges an admission fee of $9.95 for adults to take in five 'biomes' depicting various aspects of California's natural environment: Redwoods, sierras, deserts, coasts and valleys. These exhibits are inhabited by 160 wild animals, across 60 distinct species, including snakes, bobcats, scorpions, jelly fish and porcupines. Guests begin their journey with a motion-based attraction, called the Wild Ride Theater, that lets them experience the world through the eyes of various animals – moving like a mountain lion, buzzing like a bee – and then tour live animal exhibits and enjoy nature discussions with costumed Wilderness rangers. Of course, once guests pay to participate in the American Wilderness Experience, Ogden also makes money on the food service at its Wilderness Grill and the memorabilia at its Nature Untamed retail store. (Pine and Gilmore, 1999: 23–4)

Conclusion

The stakes are high. Should we move towards a capitalist super-nature, tuning our bodies to an economy of naturalized experiences, or to something more modest, more fluid and less market-driven? Unlike Agamben, I think there is hope, precisely born out of the heightened participations in bare life, shown up by movement, that I have tried to show up in this article. To begin with, there are the myriad activities which exist at the edge of the economic system which travel all the way from those who are simply looking for simple forms of exercise to those who are trying to sense something different. Then, there is the realm of the performance studies and arts, which, since at least the 1960s, have, through techniques as different as dance and performance art, been attempting to stimulate new corporeal sensibilities (e.g. Jones and Stephenson, 1999). And, last, there is the more general move towards a philosophy which can incorporate the body and so think thought differently (Shusterman, 1999).

Taken together, these alternative forms of biopolitics continue to allow a different time to inhabit the moment and even to flourish. Though they may be a small thing, they are not insignificant: sometimes a little can be a lot.

Notes

I would like to acknowledge the cognitive environment provided by the Scandinavian Collegium for Advanced Study in the Social Sciences (SCASSS) in the preparation of this article. Several people there discussed its contents with me and enabled me to hone the arguments, in particular Katharine Young and Barbara Kirshenblatt-Gimblett. I also very much appreciated the opportunity to present the article at the Materiality/Sociality Conference held at Brunel University. Several people at the conference provided valuable comments and especially Dede Boden, Kevin Hetherington, Bruno Latour and Steve Woolgar. Finally, the comments of the five referees helped me to further hone and clarify the article.

1. I use the word 'instincts' here to signal my intention to try to transcend humanist approaches to nature, though many of these 'instincts' are complex biological-cultural constructions.

2. Then when we say 'information' in everyday life, we spontaneously think of information as the result of a discarding of information. We do not consider the fact that there is more information in an experience than in an account of it. It is the *account* that we consider to be information. But the whole basis of such an account is information that is discarded. Only after information has been discarded can a situation become an event people can talk about. The total situation we find ourselves in at any given time is precisely one we cannot provide an account of: we can give an account of it only when it has 'collapsed' into an event through the discarding of information. (Norretranders, 1998: 109)

3. Many of these bodily practices necessarily contain improvisational elements, since they are always performative, instantiated in the capacities of particular bodies and content-specific (see Hayles, 1999; Thrift, 2000). Think only of the face with its potent muscular geography (cf. Brothers, 1997; McNeil, 1998; Taussig, 1998).

4. This is a very different notion of metaphor from that employed by Lakoff and Johnson (1998) which seems to me to over-determine both the idea of metaphor and the process of metaphorization.

5. In other words, the notion of speed is part of the rhetoric of how Euro-American societies go on.

6. There are interesting connections here with all kinds of earlier body practices from drill to dance which could be brought out and which are brought together in the 20th-century city in the work of writers like Laban (see Thrift, 2000).

7. The practice of photography, in other words, is as important for its *process of doing* as for its results (photographs which are normally rarely looked at).

8. Aided, in certain cases, by stimulants like drugs.

9. All these practices are heightened by the growing sense, stimulated by the media, of audience (see Abercrombie and Longhurst, 1998); we now constantly see and take in other body practices: ways of walking and the like. This mundane anthropology is becoming more and more important.

10. Thus, for example:

> . . . when we approach a great fir on the crest of a mountain, we stand tall or our eyes travel upward to the clouds and eagles, when we approach a willow our gaze sweeps in languid arcs across the backs of lime branches rippling over the lake. When we come upon a fallen tree, we have difficulty seeing it is a willow or a pine or a tree; it appears as a thicket about a log, in a confused lay out inviting closer scrutiny. (Lingis, 1998: 53)

11. Note also Derrida's thoughts on nature as a form of writing (see Kirby, 1997).

12. Ancient Christian prayer cited in de Certeau (1992: 1).

13. The main marketing slogan of the American Wilderness Experience.

14. So think of the following quotation as a business proposition.

> To recognise a person is to recognise a typical way of addressing tasks, of envisaging landscapes, of advancing hesitantly and cautiously or ironically, of playing exuberantly down the paths to us. Someone we know is someone we relate to posturally, someone we walk in step with, someone who maintains a certain style of positioning himself or herself and gesticulating in conversation and with whom we take up a compromising position as we talk. (Lingis, 1998: 53)

15. I realize that this section might be read as a Baudrillardian account of the rise of simulacra. This is not, however, the way I would want it read. Baudrillard's accounts are far too sweeping for me, and lack any but a stylized historical sense.

References

Abercrombie, N. and B. Longhurst (1998) *Audiences: A Sociological Theory of Performance and Imagination*. London: Sage.

Agamben, G. (1998) *Homo Sacer: Sovereign Power and Bare Life*. Stanford, CA: Stanford University Press.

Ansell Pearson, K. (1997) *Viroid Life: Perspectives on Nietzsche and the Transhuman Condition*. London: Routledge.

Ansell Pearson, K. (1999) *Germinal Life: The Difference and Repetition of Deleuze*. London: Routledge.

Asad, T. (1993) *Genealogy of Religion*. Baltimore, MD: Johns Hopkins University Press.

Bertman, S. (1998) *Hyperculture: The Human Cost of Speed*. New York: Praeger.

Brand, S. (1999) *The Clock of the Long Now*. London: Phoenix.

Brothers, L. (1997) *Friday's Footprint: How Society Shapes the Human Mind*. New York: Oxford University Press.

Burns, T. (1992) *Erving Goffman*. London: Routledge.

Cache, B. (1995) *Earth Moves: The Furnishing of Territories*. Cambridge, MA: MIT Press.

Clark, A. (1997) *Being There: Putting Brain, Body, and World Together Again*. Cambridge, MA: MIT Press.

Connolly, W.E. (1999) 'Brain Wars, Transcendental Fields and Techniques of Thought', *Radical Philosophy* 94: 19–28.
Crawshaw, C. and J. Urry (1997) 'Tourism and the Photographic Eye', pp. 176–95 in C. Rojek and J. Urry (eds) *Touring Cultures: Transformations of Travel and Theory*. London: Routledge.
Dagognet, F. (1992) *Etienne-Jules Marey: A Passion for the Trace*. New York: Zone Books.
De Certeau, M. (1992) *The Mystic Fable, vol. 1:. The Sixteenth and Seventeenth Centuries*. Chicago, IL: University of Chicago Press.
Deleuze, G. (1988) *Bergsonism*. New York: Zone Books.
Demasio, A. (1999) *The Feeling of What Happens: Body, Emotion and the Making of Consciousness*. London: Heinemann.
Dening, G. (1996) *Performances*. Chicago, IL: University of Chicago Press.
Feldenkrais, M. (1972) *Awareness through Movement: Health Exercises for Personal Growth*. New York: Harper and Row.
Finnegan, R. (1989) *The Hidden Musicians*. Cambridge: Cambridge University Press.
Flaherty, M.G. (1998) *A Watched Pot: How We Experience Time*. New York: New York University Press.
Foucault, M. (1984) 'On the Genealogy of Ethics: An Overview of Work in Progress', in P. Rabinow (ed.) *The Foucault Reader*. New York: Pantheon.
Gibson, J.T. (1966) *The Senses Considered as Perceptual Systems*. Boston, MA: Houghton Mifflin.
Giddens, A. (1991) *Modernity and Self-Identity*. Cambridge: Polity Press.
Gil, J. (1998) *Metamorphoses of the Body*. Minneapolis: University of Minnesota Press.
Gottdiener, M. (1997) *The Theming of America: Dreams, Visions and Commercial Spaces*. Boulder, CO: Westview Press.
Grosz, E. (ed.) (1999) *Becomings: Explorations in Time, Memory and Futures*. Ithaca, NY: Cornell University Press.
Hall, E.T. (1990) *The Hidden Dimension*. New York: Anchor Doubleday.
Hartley, L. (1995) *Wisdom of the Body Moving: An Introduction to Body–Mind Centring*. Berkeley, CA: North Atlantic Books.
Hayles, N.K. (1999) *How We Became Posthuman: Virtual Bodies in Cybernetics, Literature and Informatics*. Chicago, IL: University of Chicago Press.
Hillman, D. and D. Massio (1997) *The Body in Parts: Fantasies of Corporeality in Early Modern Europe*. New York: Routledge.
Hughes-Freeland, F. (ed.) (1998) *Ritual, Performance, Media*. London: Routledge.
Hutchins, E. (1995) *Cognition in the Wild*. Cambridge, MA: MIT Press.
Jones, A. and A. Stephenson (eds) (1999) *Performing the Body, Performing the Text*. London: Routledge.
Jordanova, L. (1994) 'The Hand', pp. 252–9 in L. Taylor (ed.) *Visualizing Theory*. New York: Routledge.
Kern, S. (1983) *The Culture of Space and Time, 1880–1914*. Berkeley: University of California Press.
Kirby, V. (1997) *Thinking Flesh*. London: Routledge.
Kovach, B. and T. Rosenstiel (1999) *Warp Speed: America in the Age of Mixed Media*. New York: Century Press.
Lakoff, G. and M. Johnson (1998) *Philosophy in the Flesh: The Embodied Mind and its Challenge to Western Thought*. New York: Basic Books.
Latour, B. and E. Hermant (1998) *Paris Ville Invisible*. Paris: La Decouverte/Institut Synthelabo.
Lingis, A. (1998) *The Imperative*. Bloomington: Indiana University Press.
Lowen, A. (1975) *Bioenergetics*. Harmondsworth: Penguin/Arkana.
McCrone, J. (1999) *Going Inside: A Tour Round a Single Moment of Consciousness*. London: Faber and Faber.
MacDonald, G. (1998) *The Complete Illustrated Guide to the Alexander Technique*. Shaftesbury, Dorset: Element Books.

McGowan, D. (1997a) *Alexander Technique: Original Writings of F.M. Alexander*. Burdett, NY: Larsons.
McGowan, D. (1997b) *Constructive Awareness: Alexander Technique and the Spiritual Quest*. Burdett, NY: Larsons.
McNeil, D. (1998) *The Face*. London: Hamish Hamilton.
Malbon, B. (1999) *Clubbing*. London: Routledge.
Margulis, L. (1998) *The Symbiotic Planet*. London: Weidenfeld and Nicolson.
Margulis, L., D. Sagan and W. Schwartz (1999) *Slanted Truths: Essays on Gaia, Symbiosis and Evolution*. Berlin: Springer Verlag.
Massumi, M. (1996) 'The Autonomy of Affect', pp. 217–39 in P. Patton (ed.) *Deleuze: A Cultural Reader*. Oxford: Blackwell.
Mattelart, A. (1996) *The Invention of Communication*. Minneapolis: University of Minnesota Press.
Muecke, S. (1999) 'Travelling the Subterranean River of Blood: Philosophy and Magic in Cultural Studies', *Cultural Studies* 13: 1–17.
Norretranders, T. (1998) *The User Illusion: Cutting Consciousness Down to Size*. New York: Viking.
Pine, J. and J.H. Gilmore (1999) *The Experience Economy*. Boston, MA: Harvard Business School Press.
Rajchman, J. (1998) *Constructions*. Cambridge, MA: MIT Press.
Reason, J.T. (1982) 'Sensory Processes', pp. 218–52 in A. Taylor (ed.) *Introducing Psychology*. Harmondsworth: Penguin.
Rival, L. (ed.) (1998) *The Social Life of Trees: Anthropological Perspectives on Tree Symbolism*. Oxford: Berg.
Roth, G. (1989) *Maps to Ecstasy*. London: HarperCollins.
Roth, G. (1998) *Sweat Your Prayers: Movement as Spiritual Practice*. London: HarperCollins.
Schechner, R. (1995) *The Future of Ritual: Writings on Culture and Performance*. New York: Routledge.
Segel, H.B. (1998) *Modernism and the Physical Imperative*. Baltimore, MD: Johns Hopkins University Press.
Serres, M. (1995) *Angels: A Modern Myth*. Paris: Flammarion.
Shusterman, R. (1999) 'Somaesthetics: A Disciplinary Proposal', *Journal of Aesthetics and Art Criticism* 57: 299–313.
Speak, D. (1999) *The End of Patience*. Bloomington: Indiana University Press.
Stewart, N. (1998) 'Relanguaging the Body: Phenomenological Descriptions and the Body Image', *Performance Research* 3: 42–58.
Sudnow, D. (1993) *Ways of the Hand*. Cambridge, MA: MIT Press.
Taussig, M. (1994) 'Physiognomic Aspects of Visual Words', pp. 205–12 in L. Taylor (ed.) *Visualizing Theory*. New York: Routledge.
Taussig, M. (1998) 'Crossing the Face', pp. 224–43 in P. Spyer (ed.) *Border Fetishisms: Material Objects in Unstable Spaces*. New York: Routledge.
Thrift, N.J. (1995) 'A Hyperactive World', in R.J. Johnston, P.J. Taylor and M. Watts (eds) *Geographies of Global Transformation*. Oxford: Blackwell.
Thrift, N.J. (1996) *Spatial Formations*. London: Sage.
Thrift, N.J. (1997) 'The Still Point: Resistance, Expressive Embodiment and Dance', in S. Pile and M. Keith (eds) *Geographies of Resistance*. London: Routledge.
Thrift, N.J. (1998) 'Steps to an Ecology of Place', in D. Massey, J. Allen and P. Sarre (eds) *Human Geography Today*. Cambridge: Polity Press.
Thrift, N.J. (2000) 'Afterwords', *Environment and Planning D. Society and Space* 19(2): 213–55.
Wallace, A.C. (1993) *Walking, Literature, and English Culture: The Origins and Uses of Peripatetic in the Nineteenth Century*. Oxford: Clarendon Press.
Walton, K.L. (1990) *Mimesis as Make-Believe: On the Foundations of the Representational Arts*. Cambridge, MA: Harvard University Press.

Watson, S. (1998a) 'The Neurobiology of Sorcery: Deleuze and Guattari's Brain', *Biology and Society*: 4.

Watson, S. (1998b) 'The New Bergsonism', *Radical Philosophy* 92: 1–23.

Weiss, B. (1996) *The Making and Unmaking of the Haya Lived World: Consumption, Commodification and Everyday Practice*. Durham, NC: Duke University Press.

Wilson, F.R. (1998) *The Hand: How its Use Shapes the Brain, Language and Culture*. New York: Pantheon.

Nigel Thrift is a Professor of Geography at the University of Bristol. His publications include *Spatial Formations* (1996), *Money/Space* (with Andrew Leyshon, 1997), *Shopping, Place and Identity* (with Danny Miller, Peter Jackson, Beverley Holbrook and Mike Rowlands, 1998), *Thinking Space* (co-edited with Mike Crang, 2000) and *City A–Z* (co-edited with Steve Pile, 2000).

The Climbing Body, Nature and the Experience of Modernity

NEIL LEWIS

> I wish to speak to the despisers of the body. Let them not learn differently nor teach differently, but only bid farewell to their own bodies – and so become dumb. (Nietzsche, 1969: 61)

What follows is an interpretation, the author's 'perspective seeing' as Nietzsche would put it, of the British tradition of adventure rock-climbing. The article forms part of an on-going project on British rock-climbing and modernity, and delineates the possibilities for a sensuous appreciation of both the human body and the physical world. As a distinctive cultural practice of the last 200 years, British rock-climbing provides fertile ground for sociological investigations into such contemporary issues as human embodied agency, sensory awareness, spatiality and, more broadly, the phenomenalities of nature and modernity. Adventure climbing is utilized as an analytical tool in order to convey the 'radical materiality' of human existence, and our overwhelmingly tactile engagement with the phenomenal world. In doing so, one finds the ground of 'another modernity' (see Lash, 1999).

Adventure climbing is presented as an act of intentionality that thwarts the desensitizing and pacifying proclivity of the body under modernity, providing a moment of remembrance, a time of reaffirmation to that which modernity ultimately seeks to replace: human embodied agency. In the perceptual world of the senses, the *real* world of the adventure climber is mapped via tactile navigation, via a body moving through an environment. Consequently, freedom becomes a form of embodied awareness: a choosing to sense and, more specifically, a choosing to *feel* and *touch* an environment. The transformation of sensuous reality is, in many respects, to apprehend reality itself, making it available for contestation

Body & Society © 2000 SAGE Publications (London, Thousand Oaks and New Delhi),
Vol. 6(3–4): 58–80
[1357–034X(200009/12)6:3–4;58–80;015471]

and redefinition (Rodaway, 1994: 145). Put more aggressively, it is to exert an existential freedom: for *how we choose to make sense of the world significantly constitutes its reality*. To engage with the world tactually is to situate oneself consciously in that world and to have a potentially unmediated relationship with it. Modernity, on the other hand, seeks to mediate our relationship with the world. It is in this sense that modernity is conceived of as a body-centred battle regarding human embodied agency.

The body of the adventure climber, what I call the *climbing body*, physicalizes or embodies a set of ideas and practices that often belie the deep-rooted ambivalence held by modernity towards the body and sensuous knowledges. The following dualisms outline some of the key differences between the climbing body and the body created under modernity – what I call the metropolitan body. Such dualisms are best thought of as a list of bearings or orientational devices that serve to reference and guide the present study:

CLIMBING BODY	METROPOLITAN BODY
organic	inorganic
self-determined	passive
tactile	ocular
of the ground	groundless

While discussion of the metropolitan body is specific to the next section but one, it nevertheless impregnates the entire article as an implicit tension wrought of a suspicion, a thought for what the climbing body *is not*. Like all binary oppositions, the meaning of one is defined through the other. Thus the climbing body, as I delineate it, can only exist in the co-presence of the metropolitan body. It exists *because* of the metropolitan body. This co-presence is not just figurative, it is phenomenologically real, as both the climbing body and the metropolitan body are part of the 'multiplicity' of contemporary human being.

The Adventure Climber, Resistance and Modernity

> . . . the most intense and productive life of culture takes place on the boundaries. (Bakhtin, 1986: 2)

In *The Sacred Canopy* (1967), Peter Berger discusses the role played by 'marginal situations' in the construction of social reality. Often characterized as a bodily state when the individual confronts death, the marginal situation 'radically challenges *all* socially objectivated definitions of reality – of the world, of others, and of self' (1967: 44). As a moment of doubt or as the realization of one's own (and others') mortality, the marginal situation has the potential fundamentally to alter our perception of reality, of the world as it is configured from day to day. Berger states that the 'confrontation with death (be it through actually witnessing the death of others or anticipating one's own death in the imagination) constitutes what is probably the most important marginal situation' (1967: 43–4). To what extent does death continue to proffer a marginal situation under modernity?

One might begin with a consideration of the individual and collective experience of death and mortality in modern Western society. In recent years, Bauman (1992a, 1998) has persuasively argued that death is no longer a part of modern life. For Bauman, the endemic incursions of scientific knowledge and professional 'expert' systems have accentuated the distance between everyday life and encounters with death. Death has become isolated, abstracted from the flow of human experience and transformed into just another 'problem' to be fitted into our busy daily schedules (Bauman, 1998: 225). This medicalization of death removes it from the public domain; rarely is it discussed and hardly ever is it seen. As Elias (1985) laments, 'Never before have people died as noiselessly and hygienically as today . . . and never in social conditions so much fostering solitude' (1985: 85). The point is this, perhaps medicalization has muted the continuing possibility of death to realize itself fully as a marginal situation? That death has been made *un*real, an event of extra-ordinary 'bad luck' instead of inevitable and universal.

The notion that death is not here but somewhere else can hold true for much of our lives. Death can now be deferred or postponed via body projects (Bauman, 1992b; Shilling, 1993). We can hide from it through participation in 'safe' activities, numerous other interests and concerns are able to distract us from actually thinking about it, and when all else fails, experts are able to manipulate our bodies via plastic surgery, thus giving the appearance of immortality. It seems that modernity has no desire for death; that modernity seeks to banish death, to save us from, or indeed, to deny us the experience of death. Shilling (1993) goes so far as to suggest that the increasing disappearance of death from everyday life has necessitated a need to develop an industry 'for representations of death. . . . It is as if the growing removal of actual death from the public sphere has stimulated a demand for an anaesthetised "knowledge" of death' (1993: 190). Violent films, news, documentaries, emergency service dramas, and 'real life' TV where *real people really die*, perhaps not only serve as an 'empty testament to our continued existence' as Shilling concludes, but signify a desire to re-incorporate death in our daily lives.

What purpose might this serve? Why, so to speak, bring death back to life? A possible answer lies in its propensity to confer a marginal situation, to open up new possibilities for existence, or to critique the 'cognitive and normative operating procedures' of contemporary life (Berger, 1967: 24). Marginal situations open up hitherto unexpected ways to live a life; perhaps a better life can be glimpsed. To experience the marginal situation of death, however, not only confers the possibility of other existences that one might live, it also compels us to reflect upon our own life – and its finitude. Death ruptures that sense of permanence and continuity we strive for through 'survival strategies' (Bauman, 1992b), forcing us to extract our heads from the sand and face the fate of our own mortality. In doing so, in confronting our own death and that of others, we are prompted, nay, condemned to consider the mutability of human being – our embodied awareness of our ceaseless change towards death (see Dollimore, 1998). And to consider the mutability of human being is to be compelled to comprehend our organic nature: to experience the body as it ages, as it sways from moments of illness to fitness and back again, as it desires and is desired, as it rejects and is itself rejected, as it is satiated through pain and pleasure.

At the risk of sensationalizing the activity, adventure rock-climbing embodies both the possibility of witnessing the death of others and anticipating one's own death – for real and in the imagination. The traditional style of rock-climbing in the British Isles has been, and continues to be overwhelmingly, a style known as *adventure* climbing or 'free' and 'clean' climbing (see Wilson, 1998 for a useful overview). In most other climbing regions of the world different styles of climbing have predominated, such as 'aid' or artificial climbing and *sport* climbing. In his discussion of rationalization and commodification within British rock-climbing, Ian Heywood (1994) succinctly delineates some of the major differences between adventure and sport climbing:

> In adventure climbing the climbing team (usually two or three) starts from the ground, without much in the way of preliminary inspection, and ascends to the top relying on a brief guidebook description, direct observation and experience, and protecting itself with ropes and leader-placed, removable devices which do not damage the rock surface. In high standard sport climbing generally short (one pitch) routes are 'worked', perhaps over days or weeks, with the climber repeatedly resting or falling onto the frequent bolts or pitons which provide security; in most sport climbs the climber is very unlikely to suffer serious injury however many times he or she falls, jumps, or rests on the protection. The same could not be said for many adventure climbs. (1994: 185)

The adventurous nature of British rock-climbing has been lauded by some non-indigenes and castigated by others. Doug Robinson (1996: 115) writes of the defining moments when the British practice of 'clean climbing' was incorporated on to American cliffs with such reverential enthusiasm it would make most British climbers blush. For Robinson, to climb a route *cleanly* transforms not only the character of the climb (1996: 217) but also the climber's body movement during the climb. Criticizing direct and intensive aid-climbing[1] as 'not nearly kinaesthetic enough to satisfy the sense we have come to feel of our-bodies-moving-in-the-mountains', Robinson describes adventure climbing as 'climbing the rock without changing it a step, a step closer to organic climbing for the natural man' (1996: 135, 119). For my own purposes I would re-phrase this as *natural climbing for organic man*: through naturalistic climbing, through letting nature dictate the path or route of the climb, and preserving this superiority, we become most aware of our organicity, our tactile and finite presence in the world.

Paolo Vitali (1997), on the other hand, inquires of the British climbing community:

> I have never climbed in the UK but I have read certain articles indicating a consensus opinion that routes, both long and short, should be left in their original state. I have to say that your persistence, in the UK, in refusing any kind of sure and lasting protection baffles me! After all, you no longer climb with hemp ropes. . . . Since climbing styles inevitably evolve over time, I find your continued resistance to bolts perplexing. (1997: 181)

What Vitali finds unintelligible is the adventure climber's apparent scant regard 'for sure and lasting protection', for bolts. In refusing bolts, the adventure climber keeps open a far greater possibility for death. The placing of protection on an adventure climb is ultimately dependent upon the climber's own ingenuity and expertise. Fissures and points of weakness in the rock vary enormously and even when protection is possible, it will never be possible to mass-produce equipment that will fit each and every crack on adventure rock-climbs. Therefore, if the route doesn't provide the climber with points of weakness within which protection can be placed, then they climb unprotected. The protection technology of the adventure climber is designed with the sole purpose of utilizing the natural characteristics given on each climb. If there are no natural fissures or points of weakness in the rock, then there is no protection. Nature dictates. An example will illustrate what this means in practice.

This may sound obvious, but adventure climbing is about seeking as much as possible to avoid falling. To fall brings into question one's protection: how good is it? And this is a question that can only be answered in the affirmative when a fall has been held. The uncertainty of a fall is brought compellingly home during the ascent of the Scoop at Sron Ulladale in the Outer Hebrides by Paul Pritchard and Johnny Dawes:

> The mood had now become very sombre and few directions were vocalised as Johnny lurched across the overhang, pinching and struggling to keep his feet in contact with the lichenous rock. . . . Then, as he was futilely trying to place a blade in a blank seam, the piece stripped and Johnny plummeted. He disappeared under a band of overhangs and began a long pendulum. I held the ropes tight and watched with horror as they slid along the sharp lip, spraying a cloud of nylon fluff into the air. As one rope snapped, the blade tinkled on the rocks 700 feet below.

> Johnny prusiked up and, silent and worried, we did the space abseil and tried to get our heads together.
>
> After a solemn discussion we both agreed things were getting a little out of hand. Johnny was shaken up and felt unsure about going back up. (Pritchard, 1997: 56)

To fall is to ask big and serious questions concerning one's ability to protect a climb: will my protection hold? What if the rope gets shorn? What if there's rockfall? What if I . . . ? These are all real possibilities if one were to fall. Of course, it is the ambiguity of uncertainty conflated during a fall that presses against the chest of the contemporary adventure climber. It is when the presence of death is felt to be most acute. In the past, the scenario was even more bleak: to fall usually meant one off all off, the falling climber dragging off his companions (as was the case with the first ascent of the Matterhorn in 1865 and the 1903 tragedy on Scafell when four climbers all on the one rope fell to their deaths) (see Bartlett, 1993; Hankinson, 1988; Withey, 1997). This stark reality gave rise to the 'sacred doctrine that *a leader should never fall*, since it would almost always have resulted in death' (Jones and Milburn, 1988: 240; emphasis added). Furthermore, a climber should always be able to down-climb anything he or she has climbed up. In short, they should always be able to rectify a dangerous situation, to get out of jail if need be. This sacred doctrine – a most subjective form of protection, of self-discipline – carries little value for the sport climber, who climbs in the knowledge that one can fall safely. As one climber informed me: 'I've tried bolts but it wasn't the same. You could try things *you knew you couldn't really climb*.'

While the 'sacred doctrine' is no longer necessary to protect the climber, the use and availability of modern technology continue to leave open the anticipation of one's own death both actually and in the imagination. Doubt, contingency and risk remain in the repertoire of the modern adventure climber's imagination and constitute key ingredients in the recipe for adventure climbing. This has prompted Ian Heywood (1994) to conclude that 'there is about the notion of adventure climbing something fundamentally at odds with the outlook and values belonging to the process of rationalization as it has been understood and described by writers from Weber onwards' (1994: 185).

If adventure climbing is considered as a marginal situation: that via its courtship with death it has the ability to disclose alternative realities, then perhaps it is time to conjoin the pursuit of happiness, the search for personal well-being, with various forms of politically motivated behaviour. The two need not be diametrically opposed activities as Robin Jarvis (1997) imaginatively outlines in his book *Romantic Writing and Pedestrian Travel*. For Jarvis, the apparently mundane activity of walking took on a political hue during the final decades of the 18th century. Walking[2] became embodied as a form of radical politics: a symbolic gesture of egalitarianism whereby the (primarily) disillusioned middle-class undergraduate could put himself on a level with the poor and disenfranchised through attire (dress) and movement (walking). Jarvis believes that pedestrian travel was utilized as a rhetoric of poverty (1997: 10, 33–9) and as a form of social observation: a means by which data could be collected about social life (1997: 43–8).

The politicization of early walkers and Romantics such as Coleridge and Wordsworth has been developed by Regina Hewitt (1997) in her book *The Possibilities of Society*. Hewitt seeks to posit the English Romantics as harbingers of the pioneering of sociology that emerged in continental Europe during the second half of the 19th century. She skilfully delineates how Wordsworth and Coleridge 'work with a sociological viewpoint and an ideal-typical method comparable to those of the founders' (1997: xii). Walking may have been an enjoyable exertion for Wordsworth and Coleridge, however it also set them free intellectually: 'walking signifies the restlessness, negatively the uprootedness and political drivenness, more positively the *mobility*, of the radical mind' (Robinson, 1989: 52, cited in Jarvis, 1997: 33). In short, walking had become a weapon of resistance. It physicalized or embodied a critique of society through body movement as a fundamental expression of agency. In 1791, the appropriately named A. Walker was compelled to write: 'there is but one way of Travelling more pleasant than riding on horseback, and that is on foot; for then I can turn to the right and to the left' (cited Jarvis, 1997: 9). This is such a simple statement yet so powerful in conveying the desire to be free from constraint, to be able to 'turn to the right and to the left'. To move in any direction one so desired. For the walkers of the late 18th century, peripatetic movement engendered independence and self-determination. It embodied the free and radical mind. Walking was freedom (of movement) par excellence.

It is important to remember that by the end of the 18th century, walking as a means of transportation was no longer a necessary requirement for the middle- and upper-class body. For example, the use of the stagecoach for passenger travel increased eight-fold between 1790 and 1836 (Jarvis, 1997: 20). Here begins the trajectory of a particular form of incarceration: the train, the car, the aeroplane and the package-holiday merely bring us up to date with our sedentary nomadism. Wordsworth was an early critic of the package-holiday and its propensity to strip the human body of agency. He was dismissive of the large-scale organized outings that brought the labourers of the industrial towns and cities of northern England to the Lake District on day-trips. During his letter crusade against the Kendal and Windermere Railway in 1844, he wrote:

> Packing off men after this fashion, for holiday entertainment, is, in fact, treating them like children. They go at the will of their master, and must return at the same, or they will be dealt

> with as transgressors. . . . Let [the Master-manufacturers] consent to a Ten Hours' Bill, with little or, if possible, no diminution of wages, and the necessaries of life being more easily procured, the mind will develop itself accordingly, and each individual would be more at liberty to make at his own cost excursions in any direction which might be most inviting to him. (cited in Bate, 1991: 50)

It is quite surprising how appropriate Wordsworth's letter is for contemporary discussions of human agency and the leisure industry. The piece is significant for its judicious attack on the policing and control of the day-tripper, the rigidity of the holiday itinerary and the need to adhere fastidiously to timetables and such like. One can surmise that Wordsworth senses the (later) Weberian fear of bureaucratic intervention, the loss of autonomy brought about by transport innovations (the train or plane never waits!) and the diminishing prospect of the traveller having the 'liberty to make at his own cost excursions in any direction which might be most inviting to him.' Yet, of course, outside of this trajectory of incarceratory leisure – perceived by Wordsworth and contemporary social theorists such as Ritzer and Liska (1997) – there remain the walker and climber who understand their radical gesture as a performance, an expression of human embodied agency.

As an extreme experience, climbing becomes a kind of corporeal subversive politics ripe with possibility for renewal that feeds back into private and social life, inflecting it with new horizons for human embodied agency. The idea of a corporeal aesthetics of extreme experience as the basis for a subversive politics centred upon conflict and limits is explored by Neveldine (1998). His argument, drawn from romanticism and postmodernism, asks: 'How has the body . . . placed itself at risk, such that it has questioned dominant notions of what it is to be a human subject in Western society . . . ?' (1998: xviii). I would offer him the example of the body placing itself at risk through adventure climbing as a possible answer.

The Metropolitan Body

Before considering the *climbing body* in detail it may be worthwhile to consider the body at the other end of the spectrum, what we may call the 'metropolitan body'. The work of Georg Simmel is instructive here. What Simmel's work on the body may lack in volume, it makes up for in apposite relevance with regard to modern life. Simmel was perhaps the first sociologist to develop ideas regarding modernity as posing a potentially irredeemable rupture for embodied experience: that a potentially adventurous and corporeal body would coalesce into the 'average sensibility' of modern life (1997 [1991]: 219). Simmel was acutely aware of the impact modern life was having upon the body, and how physiologically it was being undermined, superseded and inscribed by technology. The key text for developing a metropolitan body is probably the lecture on 'The Metropolis and Mental Life', written in 1903, which revealed, among other things, how metropolitan life, as the epitome of modernity (Frisby, 1997: 12), appeared to pose fresh questions concerning human embodied agency. How we barely lift a finger yet seem to live life at high speed.[3] Simmel begins:

> The deepest problems of modern life derive from the claim of the individual to preserve the autonomy and individuality of his existence in the face of overwhelming social forces, of historical heritage, of external culture, and of the technique of life. The fight with nature which primitive man has to wage for his *bodily* existence attains in this modern form its latest transformation. (1997 [1950]: 174–5)

There are two interrelated points I wish to draw from the lecture. First, Simmel warns of a society of mass consumption detrimental to individual life, with the creation of false or artificial needs that serve to distance or alienate the individual from the actuality of existence. Simmel's metropolis appears as some kind of alien environment where we sense and feel the continual bombardment of things, of objects that just won't leave us alone. He claims there is an '*intensification of nervous stimulation*' as our senses become besieged by the information overload endemic to modern life. Our way of coping with this 'neurasthenia', this 'tightening of the nerves' (Rojek, 1995: 109) is to distance ourselves from the world by assuming an 'intellectual' or cognitive sensibility; the 'psychic prerogative' employed to safeguard subjective life in the metropolitan world. The intellectual sensibility or metropolitan mind is interpreted by Simmel to be a way of perceiving the world through 'that organ which is least sensitive and quite remote from the depth of the personality' (1997 [1950]: 176). The embodied implications of the statement suggest that modern individuals housed inside metropolitan bodies are becoming desensitized or indifferent to the world around them. Indeed, Simmel would later write that:

> ... the perceptual acuity of all the senses evidently sinks as culture becomes more refined.... The modern person is shocked by innumerable things, and innumerable things appear intolerable to their senses which less differentiated, more robust modes of feeling would tolerate without any such reaction. (1997: 118)

Second, and on top of this decisive shift away from corporeality and embodied awareness, Simmel intimates that metropolitan life somewhat paradoxically enslaves or constrains the individual:

> The individual has become a mere cog in an enormous organisation of things and powers which tear from his hands all progress, spirituality, and value in order to transform them from their subjective form into the form of a purely objective life ... life is made infinitely easy for the

> personality in that stimulations, interests, uses of time and consciousness are offered to it from all sides. They carry the person as if in a stream, and one needs hardly to swim for oneself. (1997 [1950]: 184)

Life becomes *easy* in the metropolis because one is no longer at the helm, making decisions. No physical effort is required for we are now being carried by the modern tide. Self-propulsion, self-determination, freedom even, have all been left behind or lost. Mental and physical exertion has been superseded by an all-pervading passivity. Life has become automatic, as it no longer requires direct human actuation. In short, our modernity of ever-increasing choice and possibilities may be nothing more than a masquerade of enslavement to commodity capitalism: 'the individual has become a mere cog'.

With regard to the body, Simmel's compelling phrase 'powers which tear from his hands all progress, spirituality, and value' evokes emphatically the enfeebling nature of modern life, and is suggestive of a dumbing-down of the senses. Simmel's reference to the hands being stripped of their use finds consonance with Norbert Elias. As part of his magnum opus *The Civilizing Process* (1994 [1939]), Elias charts the demise of the sense of touch, the unmediated contact, and the corollary redundancy of the hands in modern life. Elias quotes from the 1774 edition of La Salle's *Civilité* to illustrate the stifling cerebral and ocular tendencies of the civilizing process: 'Children like to touch clothes and other things that please them with their hands. This urge must be corrected, and they must be taught to touch all they see only with their eyes' (1994 [1939]: 166). For Elias, the increasingly visual orientation towards the world instils the desire for passive pleasures, the desire to spectate rather than to participate. Simmel too, would appear to agree that modern life is overwhelmingly ocular-centric and voyeuristic. To live in a world where one has to touch with one's eyes as opposed to with one's hands emphasized, for Merleau-Ponty (1962: 315–17), the disembodied trajectory of modernity; one that led to an even greater objectification of the world.

A few years after writing 'The Metropolis and Mental Life', Simmel appeared to point out that such a life of mental bombardment and passive bodies need not necessarily be lived. There could be found aspects of life (or more precisely, leisure) that could elude the grasp of modern consumer capitalism. One such ideal type was the adventure; an activity Simmel felt was emblematic of meaningful and valuable leisure. Written in 1910, 'The Adventure' (1997 [1958]) expressed Simmel's deep desire that there could be moments of life within which fundamental categories of existence were encapsulated or synthesized. Speaking of mountain climbing, Simmel had this to say:

> ... as a means for momentary enjoyment, which comes from the exertion of all one's energies, from playing with danger and the emotion of the panoramic view. Indeed, I would place this

> enjoyment as the highest that life can offer. The less settled, less certain and less free from contradiction modern existence is the more passionately we desire the heights. (1997 [1991]: 220)

Simmel felt that too much of modern life was not the life we wished to be living (though he was always dazzled by metropolitan living). That modern life had somehow infected the body, creating the 'average sensibility'. In modern society, 'The Faustian wish, "I stand before you, nature, a solitary individual" is evermore rarely realized and so increasingly rarely declared', he laments (1997 [1991]: 219). Simmel believed that the changes taking place in our cultural and economic life were stripping us of a particular kind of freedom – the sense of being a fully responsible agent. Of being *authentic*.[4] We feel invaded. Never fully responsible for our actions. Simmel would call this an 'increase in nervous life' (see Frisby, 1989: 80). A stress that consumes the body; a kind of tension upon the nerves. An existentialist problem: this life does not satisfy me, something is missing. And yet I *know* this. So what am I going to do about it? This is the issue I feel Simmel tried to get to grips with in 'The Adventure'. Simmel was attempting to convey a technique of life whereby the alienation of modern life could be circumnavigated, however momentarily. As an 'island in life' the adventure enabled one to escape a *blasé* attitude to all sensations; our faculties, dumbed-down by the process of modern living, could be rejuvenated or their capacities re-learned. As a state of consciousness, adventure was most likely achieved during a period of bodily arousal, a '*form of experiencing*' where 'fundamental categories of life are synthesized'; an embodied spatial and temporal moment when we take control of our lives and 'forcibly pull the world into ourselves' (1997 [1958]: 225).

Estrangement from the world is to be expected when one is already estranged from one's body. Adventurous activity, on the other hand, serves to unite body and world. Hence the fundamental flaw in dualistic approaches to the world. 'When Descartes lost the body he also lost the world', provocatively states Lynne Belaief (1977: 54), underscoring the disembodied nature, the *groundlessness* of Cartesian-style thought. And, one might add, the ocular-centric and cognitivist dwelling of metropolitan life. Climbing can thus be seen to offer a validation of the body at a time when the first serious incursions of modern technology were being felt; the climbing body thus developed in resistance to a burgeoning metropolitan body.

The Climbing Body

The body is our most fundamental means of communication and interaction with the world. We make sense of the world by acquiring information through our bodies. Merleau-Ponty emphasizes the primacy of the body in relating the human to her world and its ability to convey meaning:

> My body is the seat or rather the very actuality of the phenomenon of expression . . . [It] is the fabric into which all objects are woven, and it is, at least in relation to the perceived world, the general instrument of my 'comprehension'. (1962: 235)

The embodied character of the adventure necessitates, demands even, an active body – a body in motion. A body in motion is a body endowed with kinaesthesis. The word *kinaesthesis* comes from the Greek words *kinein*, meaning to move, and *aisthesia*, to perceive. Kinaesthesis is thus the sense that informs you of what your body is doing in space through the perception or sensation of movement in the joints, tendons and muscles. Significantly, and as part of the rebuttal of the overly visualist paradigm of modernity, kinaesthesis need not necessarily employ the sense of sight: for we know if our arm or leg is outstretched without requiring visual proof because our kinaesthetic sense has already 'told' us. Kinaesthesis is therefore an embodied state of awareness, and perhaps the most akin to having an adventurous experience.

The walking diaries of Samuel Taylor Coleridge offer numerous examples of kinaesthesis 'at work'. For here we locate an embodied state of awareness whereby the body in motion configures a particular experience of self and nature. Coleridge was wont to arrest his often adventurous pedestrian sojourns with realizations of the multi-perspectivity of self in nature. His notebooks are full of the changing orientations of the body as it moves though a landscape:

> . . . the foreground a sloping wood, sloping down to the River & meadows, the serpent River beyond the River & the wood meadows terminated by Melbreak walled by the Melbreak . . . close by my left hand a rocky woody Hill, & behind it, half hidden by it, the violet crag of Grasmere. . . . I climb up the woody Hill & here have gained the Crummock Water – but have lost the violet Crag. (cited in Jarvis, 1997: 136)

Here the 'situated I' is continually having to reassess itself in relation to the landscape: taking bearings, locating itself, moving on, taking new bearings, relocating itself, moving on. Any hill-walker (or phenomenologist for that matter) knows what's going on here: the world is experienced as a 360-degree sphere[5] within which is the situated I. Within this spherical world the human body spatializes itself along three directional axes: horizontal (forward and backwards), directional (right and left) and vertical (up and down). 'These various directional axes disclose themselves through the concrete movements of the lived body' (Schrag, 1988: 115). Therefore the body is spatially situated. Coleridge makes apparent the lack of an Archimedean viewpoint; spatially, we are always somewhere, always viewing the world from a particular place at a particular time.

There is also a suggestion of the situated I being 'in touch' with the environment around it: 'close by my left hand a rocky woody Hill . . .'; that the kinaesthetic sense enables us to be in touch with our surroundings. For Gibson (1968), kinaesthesis is far more than just a metaphor for touch, it is one of its faculties. As Rodaway (1994) summarizes 'Touch is . . . about both an awareness of presence and of locomotion' (1994: 42). Touch, therefore, is about being aware of being somewhere. Being situated. To paraphrase Simmel, the 'technique of life' in the metropolis utilizes different sensuous knowledges to those of the adventure. While the adventure emphasizes immersion, participation and spiritual depth, metropolitan life suggests reservation, spectatorship and a spiritual life 'remote from the . . . personality'. Here, then, is an intimation of how the physical and cultural worlds become inscribed upon the body and how the body physicalizes its relationship with the world. Simmel's concluding remarks upon 'adventure' serve well as a summary of the article thus far:

> We are the adventurers of the earth; our like is crossed everywhere by the tensions which mark adventure. But only when these tensions have become so violent that they gain mastery over the material through which they realise themselves – only then does the 'adventure' arise . . . it is the radicalness through which it becomes perceptible as a life tension, as the rubato of the life process, independent of the materials and their differences. (1997 [1958]: 232)

The Hands of the Climber

In our role as 'adventurers of the earth', the hand is central. Historically, and cross-culturally, the hands have often been the bearers of symbolic value, conveyors of meaning for classificatory systems that establish criteria for social identities and codes of conduct (see Turner, 1992 for a useful overview). Not only is it the key tool of our dual self-definition as *homo faber* and *homo ludens*, as creators of new worlds and destroyers of old, and the means by which we sculpt thought, the hand is also the privileged discloser of our most intimate sensing of the world. Touch is perhaps the close contact sense par excellence for it emphasizes the 'direct experience' of the informational content of knowledge, the unmediated acquisition of embodied knowledge. Mark Goodwin (1996) eloquently expresses the centrality of the hand with direct experience within rock-climbing in his poem 'Craft':

> Pull
> up on jugs of gritstone. Crimp
> on folds of compressed sand. File
> movement smooth in mind. Touch
> rough quartz studs. Fashion
> delicate pain in ornate thoughts. Jam
> soft flesh fists tight in cracks. Sculpt
> in the medium of wind-carved flakes and overlaps.

For the rock-climber it is the hands, all fingers crimping and fists jamming, and the sense of touch, that communicate most readily the state of play of climbing. It is the hand that communicates the informational content of knowledge of a particular climb. Touch is the pivotal sense utilized by climbers in learning to know a climb.[6] Climbers *feel* their way up a route via tactile navigation. The centrality of the hands and the sense of touch to processes of receiving knowledge of the world around us is also apparent in the work of the symbolic interactionist pioneer, George Herbert Mead. In *Mind, Self, and Society* (1934) Mead writes of the hand as an intermediate point of contact between the world as 'physical thing' and the world as consumed act, or, put more appropriately, the world as cliff environment and the world as *grasp*-ed knowledge. Mead writes:

> . . . when we ordinarily speak of things about us we refer to physical things. The characters that go to make these up are primarily determined by the hand. . . . If we speak now of the animal as constituting its environment by its sensitivity, by its movements towards the objects, by its reactions, we can see that the human form constitutes its environment in terms of these physical things which are in a real sense the products of our own hands. (1934: 184–5, 249)

The hand is thus capable of being both mediator (it has the capacity to resolve or transcend subject/object dualisms such as 'man' and 'physical thing'), and conduit (hands become a means of access, of communication and knowledge), for the processing of the informational content of knowledge of the world. To paraphrase Simmel, it is via the hands that we 'pull the world into ourselves'. Moreover, it is the sensitivity of our hands that is responsible for relaying so much of our knowledge of the world around us. Tactile navigation – the kinaesthetic moving/touching of the body – is the total embodied awareness of a body in an environment. Knowledge is made corporeal with the sense of touch replacing that of sight as the primary mode of gathering data. For example, the poet, educationalist and climber Geoffrey Winthrop Young reflecting on his party's first ascent of the south face of the Täschhorn in the Swiss Alps said: 'Never can an experienced party have been more *deceived by the look* of a precipice from its foot' (Hankinson, 1995: 87; emphasis added), which suggests the paucity of vision alone when seeking to know or make sense of an environment.

Similarly, cognizance of the written word – for example, as in the textual description of the rock-climb or route in the climber's guidebook – is fundamentally undermined as being the principal means by which the climb is understood. The direct experience of climbing usurps the pre-eminence of cognitive apprehension as the key to acquiring knowledge. Instead, one can talk of an embodied knowledge of the climb through which the climber re-orientates herself with the world. This re-orientation is via embodied experience, a corporeal knowing rather than a cognitive knowing. 'Our fingers grasped, our toes levered, our joints flexed and bent, our senses smelt and saw rock. Each move was a response to the unmediated features of this part of the world's surface' (Craig, 1996: 107). In many ways, this re-orientation represents an inversion of modernity's emphasis upon the dominance of the text as *the* source of knowledge. As Mellor and Shilling (1997) argue:

> Cognitive apprehension assumes that valid knowledge is gained from mental activity freed from the bodily prejudices of emotions, and considers the mind and body to be fundamentally separate entities. . . . Somewhat ironically, cognitive apprehension presupposes that one can increase knowledge only by losing a degree of sensory and sensual contact with that knowledge. (1997: 58)

Ironic it certainly is, for it would appear that *touch* could actually increase our valid knowledge of the world. Thus contrary to the assumptions of cognitive apprehension – and by this Mellor and Shilling mean the emergence and subsequent dominance during modernity of a particular way of *mental looking* at the world (1997: 24), epitomized most emphatically by scientific rationality – a diminution in sensuous contact actually undermines our capacity to attain valid knowledge of the world. Moreover, this misplaced emphasis upon the cognitive ushers forth an existential crisis: the 'inevitable sense of personal incompleteness and the difficulties of "keeping in touch" with knowledge and meaning' when the body is estranged from the process of knowledge production (1997: 155).

Such an ethos is at complete odds with climbing culture. For here, the hands of the climber are the talismanic parts of the body. 'My fingers (mine, yes, look, they're on the ends of my hands) caress the underclings as though they were razors', exclaims David Craig (1996: 93) as if realizing for the first time what his fingers are for. Unlike the feet, wrapped in sticky rubber shoes, the climber's hands have an unmediated relationship with the natural world. The engagement between hand and rock (or nature) is thus pure and direct. 'It is not consciousness which touches or feels, but the hand', argued Merleau-Ponty (1962: 316). This is certainly the case for climbers who feel their way along a climb:

> . . . we basked in summer-like warmth that made my fingers slip as though buttered when I set off rightwards from the crag's western edge. . . . I teetered sideways sweating. Black mini-steps, rims for side-pulls on fingerends, little vertical shelves that end in mid-air and force me to invent a grip in which I cup my fingers under a shelf and pinch it before laying-away rightwards from it. . . . At last I'm striding across a steep gutter and sinking my arms behind a big flake of pale-orange Langdale rock before embracing the grey-satin trunk of a holly with utter gratitude. . . . The joint under the eave, previously closed, has relented slightly. I stick my fingerends in, palm upwards, and brace my knuckles downwards against the glacis. The tiny rim has become the hinge of my world . . . (Craig, 1996: 204–5)

Thus, above all else, it is the hand that informs us whether or not a climb feels 'right' (or climbable). Our most fundamental mode of knowledge therefore 'is

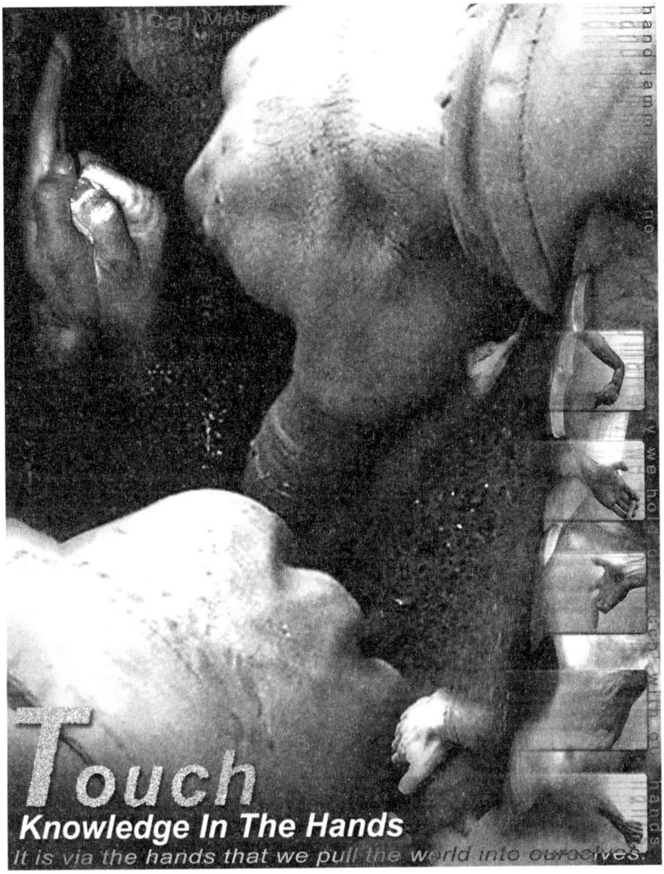

Figure 1 Hands: the talismanic zone of the climbing body (photo design by Max Bretherton).

knowledge in the hands' (Merleau-Ponty, 1962: 144). And perhaps for this reason the hands of the climber take on symbolic status. The primary mode of knowledge is gathered through the hands. Of course, there are more mundane reasons why the hands should be blessed with significance: the tensions of climbing a hard route are always eased when a good handhold is reached. All climbs become bearable if you can get a good handhold! The hand-jam is a notorious case in point (see Figure 1). Here the climber inserts a clenched hand into a crack before releasing the clench in order to let the hand mould itself to fit or 'jam' in the aforementioned crack. The climber then loads the (hand) jam with body weight in order raise the feet. A difficult and much admired climbing technique, and one which, for many, feels decidedly unnatural. This is hardly surprising, for hand-jamming is not the normal way we hold or 'touch' with our hands.

Inscribing the Body

Over time, the practice of rock-climbing and the frequent and sometimes distinctive way the climber utilizes her body begins to take on an embodied form. Through the cuts and abrasions, the freezing cold and sun traps, the taut muscles and creaky joints, the practice of climbing inscribes itself upon the body and, most emphatically, upon the hands. But touching is a reciprocal act: to touch something is to let something touch you in return (Montagu, 1971; Rodaway, 1994: 41). After inscribing with his body a new climb at Ramshaw Rocks in Staffordshire, a venue notorious for its coarse rock, Joe Brown 'lay gasping at the top with the blood pouring from the tattered strips of flesh which before had been the backs of his hands' (Jones, 1997 [1973–4]: 103). In mapping the rock with his body – fingers, fists, toes, feet, blood, sweat, obscenities – Brown is, in turn, mapped by the rock. For climber and environment inscribe each other.[7]

The climbing body is worked upon through the very act of climbing: it is recast, moulded and shaped, transformed and, in substance, created through the act of climbing and embodied engagement or immersion with rock. The practice of rock-climbing trains or cultivates the body towards a better configuration for climbing. The phrase 'the best training for climbing is climbing', a well-rehearsed truism within climbing culture, indicates the centrality of the actual practice of climbing for the formation of climbing bodies – as opposed, for example, to developing a climbing body by pulling weights in a gymnasium. To paraphrase Marcel Mauss (1973 [1934]), the 'techniques of the body' most attuned to climbing are discovered and created through the very act of climbing itself. Winthrop Young expressed this realization – that one is somehow miraculously endowed with a climbing body as a consequence of climbing – with unbridled joy: 'I remember wondering at my body – the worn fingernails, the bruised knees, and the lump of climbing muscle that had begun to bunch above the arch of the foot, seeing it as beautiful in relation to this new purpose' (cited in Craig, 1996: 154).

Thus the body has the propensity to physicalize and convey its own sensibility, to become a matrix of, and for, inscription. Such a standpoint of conscious mutability highlights the body-centred battle within modernity, that the body feels changes in both its material and ideational environments. Throughout his career, Michel Foucault continually made recourse to the human body as 'the inscribed surface of events' (1977a: 148), the body as the template of culture, noting how societies 'invest it, mark it, train it, torture it, force it to carry out tasks, to perform ceremonies, to emit signs' (1977b: 25). Taking Nietzsche as his inspiration, Foucault argues that the path towards freedom lies in *doing*, in activity: 'Liberty', he maintains, 'is a *practice*'; it is 'what must be exercised' (1984: 245). It is the body that must be 'exercised' first. The foundations of the self are to be laid through the 'invention' of the self, for Foucault urges us to 'create ourselves as a work of art' (1984: 351). Freedom, then, is first made corporeal. And it is in the making of the body, the struggle for self-creation, that freedom is most evidently actualized. As Leslie Paul Thiele (1990) concludes his discussion of Nietzsche's heroic individual: 'Like all heroes, he consciously writes his autobiography with his deeds' (1990: 48). The same might also be said of Foucault's self-creating individual who invents and maintains his identity through the '*stylized repetition of acts*' (Butler, 1990: 140). Again, we are emphasizing the corporeality of the inscription process. The body physicalizes discourses of how to engage (or how one would wish to engage) with the world. Let me illustrate this by returning to the hand of the climber:

> He couldn't get over my hands. He sat at his desk for three or four minutes in total silence, lightly grasping between thumb and forefinger the tips of my middle fingers and reading not the palms for my future but the backs for my past. The recent fortunes of the gritstone climber are always written there. Large brown scabs, small red beads, formed striations in all directions on a sore groundrash. Finally he asked for an explanation but seemed unable to accept it. With enthusiasm I went through the technique of hand-jamming but he was a poor visualiser. He shook his head in impatient disbelief. He thought there must be some eczema there, exacerbated perhaps by this strange activity.
>
> I was shocked. . . . To lend support to my story I showed the doctor the detail of the fingerprints. All displayed that telltale circle, about a centimetre in diameter, within which the outer skin is worn away. (When I was to resume gritstone climbing, for a spell of three years I needed only to cut my thumb nails. The rock kept the nails of the fingers trimmed to just the right length.) (Drasdo, 1997: 17–18)

This is Harold Drasdo recounting his National Service medical examination of 1947. Ironically, his body, currently displaying the disciplinary stigmata of the climbing body, is about to come under the influence of another discipline – that of military life. Drasdo's hands physicalize the discourse of climbing, and this discourse emphasizes nature and the natural. The practice of touching the rock inscribes itself upon the body. The climber's hand, so often immersed in nature, becomes imbued with natural qualities: 'The rock kept the nails of the fingers trimmed to just the right length'. The practice of rock-climbing cultivates the body towards a better configuration for climbing. And this body is, in substance, created by nature. Through its very engagement with nature, the climbing body becomes natural. It returns to nature. Climbers frequently express this sentiment of the reunion of body and nature:

> At the very instant my hands and feet came on the rock six years rolled away in a flash. The rock was not strange, but familiar. At each move I was taking the right holds at the right time – but no, I did not 'take' the holds – of their own accord they *came* to me. Hand, foot, and eye – nerve and muscle – they were co-ordinating, and my climbing was effortless. (Murray, 1951: 10)

Here Bill Murray returns to rock for the first time in six years. Due to the Second World War, much of Murray's life became an adventure: he fought in the Libyan Desert, was captured and spent three years in various central European prison camps where he wrote famously of *Mountaineering in Scotland* (1947). However, rock-climbing, for Murray, was as much of a coming-home as it was an adventure; the mountain environment 'was not strange, but familiar'. When climbing, Murray's sensuous knowledge becomes apparent and his body returns to being a climbing body, a body in nature: 'I did not "take" the holds . . . they *came* to me.' Along with generations of fellow climbers, Murray had the desire to explore the world, to embody it and to convey agency: 'it is by our physical presence on the cliffs and ridges, by the exploration of their gullies no less than by wandering the plateau . . . that we come fully to know these mountains and win the high reward of intimacy' (1947: 229). For Murray and many other climbers, it is the natural world that provides the necessary sense of belonging and locus for one's identity, for it is there that they can fully express their embodied selves. Simmel might have concurred that the techniques of life under modernity have stunted our ability to utilize sensuous knowledges and to exercise our embodied agency. Climbers too know this.

Conclusion

Throughout this article there have been three key themes united in their attempt to reveal the human body as both the first and last 'point of resistance to the global imperatives' of the modern age (Mirzoeff, 1995: 1). First, I emphasized the need to consider the organic nature of human being, to consider the marginal situation of death and our consequent sense of mutability as providing significant moments to remember the body. In doing so, a key impulse of modernity is usurped: the increasing denial of the body's organicity. Our mutation unto death. Second, adventure climbing was put forward as an example of bodily resistance, a means by which death and mutability are kept alive via specific practices. The imposing of limits, such as the adventure climber's refusal to utilize a form of technology (the bolt) that would severely reduce the consequences of a fall, serves to keep alive the prospect of death. Third, an embodied structure of thought was developed, in order to convey the radical materiality of the world and our overwhelmingly tactile engagement with it. The 'metropolitan body' as envisaged by Simmel was set up against the climbing body in order to draw attention to the body's sense and use of perception. A phenomenological perspective attempted to glean the climbing body's primacy of touch and the use of the hands for 'looking'.

The climbing body physicalizes or embodies various ideas seemingly at odds with many of the key tropes of modernity. For unlike the prevalent 'scopic regimes' of modernity (Jay, 1992), the climbing body advocates a sensuous appreciation of both the human body and the physical world. Sensuous knowledges provide the informational content of knowledge utilized by climbers to 'make sense' of the world. As an embodied self in nature, a 'situated I', the climbing body orientates itself through tactile navigation. The climbing body locates itself by *feeling* its way through the world. Physical things are understood through the sense of touch; for the climber, knowledge is *grasp*-ed. Therefore the prevailing 'shared vocabularies of body idiom' (Goffman, 1963: 35) expressed by climbers complement a mutating and growing corporeal discourse within the human sciences (captured by journals such as *Body & Society*) by proffering a critique of the visually dominant and disembodied paradigm of modernity. Adventure climbing is, perhaps, modernity's quintessential embodied ritual of resistance.

Notes

1. 'Aid climbing' is essentially to climb rock with a panoply of technical instruments to assist the actual *climbing of rock*. In effect, the climber lays the path ahead with various implements to help facilitate his or her progress up the rock. In short, and in many instances, it is to climb without having to touch the rock with either hand or foot or both. The kinaesthetic paucity of aid climbing is given expression by the writing of Pritchard (1997). Recalling his ascent of the Central Tower of Paine in Patagonia, Pritchard writes:

> Slide jumar up as far as it will go. Inhale. Weight foot loop. Pull with right arm. Stand up straight. Exhale. Clink. Sit down in harness. Gasp. Look up. No nearer. Slide jumar. Inhale. Weight foot. Pull. Stand. Exhale. Clink. Sit. Gasp. Slide. Weight. Pull. Stand. Clink. Sit. Gasp. Slide. Weight. Pull. Stand. Clink. Sit. Gasp. Slide. Clunk. (1997: 73)

2. It is worth pointing out the affinity between walking and climbing – for the two are intimately linked with the construction of the 'climbing body'. More often than not, the two activities form part of the climber's day out. Quite simply, in order to go climbing one also has to go for a walk. As the British Mountaineering Council suggests, rock-climbing is best defined as a foot-based recreation: 'climbing, hillwalking and mountaineering are all variations of the same foot-based activity and should without doubt be regarded as "recreation on foot"' (*BMC Summit Magazine*, 1998: 4). Thus the climbing body is also a walking body.

3. When J.M.A. Thomson, a leading climber at the turn of the century, committed suicide in 1912, the coroner concluded that the increase in nervous breakdowns was 'a sign of the times. We are all going at express speed' (cited in Drasdo, 1997: 118).

4. Here I use the term 'authentic' as a sign of agency and independence, as suggested by Cooper (1990): 'for "authentic" here echoes with the meaning of the Greek word from which it derives, namely "one who does a thing himself"' (1990: 109).

5. Ingold (1993) writes of the difference between the sphere and the globe as both metaphor for the world and way of habituating the world. To conceive of the world as a globe makes it

> . . . an object of appropriation for a collective humanity . . . we do not belong to the world. . . . Rather, since our very humanity is seen to consist, in essence, in the transcendence of physical nature, it is the world that belongs to us.
>
> In the local perspective the world is a sphere . . . centred on a particular place. From this experiential centre, the attention of those who live there is drawn even deeper into the world, in the quest for knowledge and understanding. (1993: 39, 40–1)

6. But this is not to suggest that the climber is then able to *know* wholly that environment. Bill Murray's (1951) bold exclamation, 'NO MAN WILL EVER KNOW Ben Nevis. No man *has* known Ben Nevis' (1951: 164) reflects the adventure climber's proclamation that, first, the highest form of knowledge is the subjective experience of embodied knowledge, and that, second, even if she could, the climber has no desire to know nature totally. Indeed, it is the knowledge or realization of the former (that all knowledge is ultimately but a perspective however much we participate within discourse communities) that persuades the climber to adopt an adventurous attitude to the world. With too much knowledge (or knowledge of the wrong kind?) there would be no adventure, no leap into the unknown. For adventure climbers the natural world is *sublime* in the sense that it can never be contained, hermetically sealed for human consumption. Similarly, the acceptance of mutability admonishes to us: 'I too am nature. I too elude grasping. I too can never be contained.' Thus, when Simon Critchley (1997) argues that we are unable to comprehend death in a manner so as to enable it to be life-affirming, that we can never grasp meaningful existence within it (1997: 25, 75), I would want to offer the example of adventure climbing as an activity that utilizes death positively. The author's understanding of adventure climbing renders the potentiality that death, as possibility and actuality, can be harnessed as part of a moral code of practice. Adventure climbing is a discipline through which climbers apprehend the pivotal role played by their bodies in the delimitation of our natural and technological existences.

7. Lack of space curtails a thorough discussion of how the body inscribes its environment: whether through schemes of classification (the naming and recording of place), poeticization (the storytelling and mythologizing of landscape) or via embodiment (the sculpting process of hands and feet), bodies and environments inscribe each other. As a starting point consider Drasdo's (1978 [1974]) proposition:

> . . . it might be claimed that cliffs and mountains are facades without shape or dimension until they are floodlit by human effort . . . the climber's touch brings the cliff to life. . . . Climbs interpret mountain faces. A climb is the most human relationship possible with a mountain face. Climbs amplify the persona of a mountain. The more effort has been expended, the more increment to the mountain's character. (1978 [1974]: 457–8)

References

Bakhtin, M. (1986) *Speech Genres and other Late Essays*, trans. V.W. McGee. Austin: University of Texas Press.

Bartlett, P. (1993) *The Undiscovered Country: The Reason We Climb*. Glasgow: The Ernest Press.

Bate, J. (1991) *Romantic Ecology: Wordsworth and the Environmental Tradition*. London: Routledge.

Bauman, Z. (1992a) *Mortality, Immortality and Other Life Strategies*. Cambridge: Polity Press.

Bauman, Z. (1992b) 'Survival as a Social Construct', *Theory, Culture & Society* 9(1): 1–36.

Bauman, Z. (1998) 'Postmodern Adventures of Life and Death', pp. 216–31 in G. Scambler and P. Higgs (eds) *Modernity, Medicine and Health*. London: Routledge.

Belaief, L. (1977) 'Meanings of the Body', *Journal of the Philosophy of Sport* 4: 50–68.

Berger, P.L. (1967) *The Sacred Canopy: Elements of a Sociological Theory of Religion*. New York: Doubleday.

BMC Summit Magazine (1998) Issue 11. Kettering: Greenshires Publishing.

Butler, J. (1990) *Gender Trouble*. New York: Routledge.

Cooper, D.E. (1990) *Existentialism: A Reconstruction*. Oxford: Blackwell.

Craig, D. (1996) *Native Stones: A Book about Climbing*. London: Pimlico.

Critchley, S. (1997) *Very Little . . . Almost Nothing: Death, Philosophy, Literature*. London: Routledge.

Dollimore, J. (1998) *Death, Desire and Loss in Western Culture*. London: Penguin.

Drasdo, H. (1978 [1974]) 'Climbing as Art', pp.455–61 in K. Wilson (ed.) *The Games Climbers Play*. London: Bâton Wicks.

Drasdo, H. (1997) *The Ordinary Route*. Glasgow: The Ernest Press.

Elias, N. (1985) *The Loneliness of the Dying*. Oxford: Basil Blackwell.

Elias, N. (1994 [1939]) *The Civilizing Process*. Oxford: Blackwell.

Foucault, M. (1977a) *Language, Counter-Memory, Practice: Selected Essays and Interviews*. Ithaca, NY: Cornell University Press.

Foucault, M. (1977b) *Discipline and Punish: The Birth of the Prison*. London: Penguin.

Foucault, M. (1984) *The Foucault Reader*, ed. P. Rabinow. London: Penguin.

Frisby, D. (1989) 'Simmel and Leisure', pp. 75–91 in C. Rojek (ed.) *Leisure for Leisure: Critical Essays*. London: Macmillan.

Frisby, D. (1997) 'Introduction', in D. Frisby and M. Featherstone (eds) *Simmel on Culture: Selected Writings*. London: Sage.

Gibson, J. (1968) *The Senses Considered as Perceptual Systems*. London: George Allen and Unwin.

Goffman, E. (1963) *Behaviour in Public Places: Notes on the Social Organisation of Gatherings*. New York: Free Press.

Goodwin, M. (1996) 'Craft', *Climber Magazine* June (Glasgow: Caledonian Magazines Ltd).

Hankinson, A. (1988) *A Century on the Crags: The Story of Rock-Climbing in the Lake District*. London: J.M. Dent.

Hankinson, A. (1995) *Geoffrey Winthrop Young: Poet, Educator, Mountaineer*. London: Hodder and Stoughton.

Hewitt, R. (1997) *The Possibilities of Society: Wordsworth, Coleridge, and the Sociological Viewpoint of English Romanticism*. New York: State University of New York Press.

Heywood, I. (1994) 'Urgent Dreams: Climbing, Rationalization and Ambivalence', *Leisure Studies* 13: 179–94.

Ingold, T. (1993) 'Globes and Spheres: The Topology of Environmentalism', in K. Milton (ed.) *Environmentalism: The View from Anthropology*. London: Routledge.

Jarvis, R. (1997) *Romantic Writing and Pedestrian Travel*. London: Macmillan.

Jay, M. (1992) 'Scopic Regimes of Modernity', pp. 178–95 in S. Lash and J. Friedman (eds) *Modernity and Identity*. Oxford: Blackwell.

Jones, C.T. (1997 [1973–4]) 'The Late Fifties: A Personal Account', pp. 102–6 in T. Gifford (ed.) *The Climbers' Club Centenary Journal*. The Climbers' Club.

Jones, T. and G. Milburn (1988) *Cumbrian Rock: 100 Years of Climbing in the Lake District*. Glossop: Pic Publications.

Lash, S. (1999) *Another Modernity: A Different Rationality*. Oxford: Blackwell.

Mauss, M. (1973 [1934]) 'Techniques of the Body', *Economy and Society* 2: 70–88.

Mead, G.H. (1934) *Mind, Self, and Society*. Chicago, IL: University of Chicago Press.

Mellor, P.A. and C. Shilling (1997) *Re-Forming The Body: Religion, Community and Modernity*. London: Sage.

Merleau-Ponty, M. (1962) *The Phenomenology of Perception*. London: Routledge.

Mirzoeff, N. (1995) *Bodyscape: Art, Modernity and the Ideal Figure*. London: Routledge.

Montagu, A. (1971) *Touching: The Human Significance of the Skin*. New York/London: Columbia University Press.

Murray, W.H. (1947) *Mountaineering in Scotland*. London: Diadem Books.

Murray, W.H. (1951) *Undiscovered Scotland*. London: Diadem Books.

Neveldine, R.B. (1998) *Bodies at Risk: Unsafe Limits in Romanticism and Postmodernism*. Albany: State University of New York Press.

Nietzsche, F. (1969) *Thus Spoke Zarathustra*. London: Penguin.

Pritchard, P. (1997) *Deep Play: A Climber's Odyssey from Llanberis to the Big Walls*. London: Bâton Wicks.

Ritzer, G. and A. Liska (1997) '"McDisneyization" and "Post-Tourism": Complementary Perspectives on Contemporary Tourism', in C. Rojek and J. Urry (eds) *Touring Cultures: Transformations of Travel and Theory*. London: Routledge.

Robinson, J. (1989) *The Walk: Notes on a Romantic Image*. Norman and London: University of Oklahoma Press.
Robinson, D. (1996) *A Night on the Ground, a Day in the Open*. La Crescenda, CA: Mountain N' Air Books.
Rodaway, P. (1994) *Sensuous Geographies*. London: Routledge.
Rojek, C. (1995) *Decentring Leisure: Rethinking Leisure Theory*. London: Sage.
Schrag, C.O. (1988) 'The Lived Body as a Phenomenological Datum', pp. 109–18 in W.J. Morgan and K.V. Meier (eds) *Philosophic Inquiry in Sport*. Chicago, IL: Human Kinetics Publishers.
Shilling, C. (1993) *The Body and Social Theory*. London: Sage.
Simmel, G. (1997 [1950]) 'The Metropolis and Mental Life', in D. Frisby and M. Featherstone (eds) *Simmel on Culture*. London: Sage.
Simmel, G. (1997 [1958]) 'The Adventure', in D. Frisby and M. Featherstone (eds) *Simmel on Culture*. London: Sage.
Simmel, G. (1997 [1991]) 'The Alpine Journey', in D. Frisby and M. Featherstone (eds) *Simmel on Culture*. London: Sage.
Simmel, G. (1997) 'Sociology of the Senses', in D. Frisby and M. Featherstone (eds) *Simmel on Culture*. London: Sage.
Thiele, L.P. (1990) *Friedrich Nietzsche and the Politics of the Soul: A Study of Heroic Individualism*. Princeton, NJ: Princeton University Press.
Turner, B.S. (1992) *Regulating Bodies: Essays in Medical Sociology*. London: Routledge.
Vitali, P. (1997) 'Bolting in the Alpine Environment', *Alpine Journal* 102(346): 181–3.
Wilson, K. (1998) 'A Future for Traditional Values', *Alpine Journal* 103(347): 175–88.
Withey, L. (1997) *Grand Tours and Cook's Tours: A History of Leisure Travel 1750–1915*. New York: William Morrow.

Neil Lewis is a research student in the Sociology department at Lancaster University. He is in the process of writing his PhD thesis, titled 'The Adventure Climber: A Culture of Nature'. An enthusiastic climber, he has climbed extensively in the British Isles.

Walking in the British Countryside: Reflexivity, Embodied Practices and Ways to Escape

TIM EDENSOR

What could be more natural than a stroll in the countryside? The air is fresh, the body realizes its sensual capacities as it strains free from the chains of urban living, and our over-socialized identities are revealed as superficial in an epiphany of self-realization. In the past two centuries, walking has shifted from central mode of transport to leisure activity. According to the Department of Transport, it is now the most popular physical activity undertaken for pleasure, and one which is apparently increasing – in 1996, 68 percent of Britons partook, an increase of 8 percent in a decade (HMSO, 1998: 16). Of all walks, 14 percent have no other objective than being undertaken for their own sake, and one third of all trips into the countryside are to go for a walk, 45 percent of which cover over two miles (HMSO, 1998: 17). The institutionalization of walking as a social practice is reflected in the authors' advocating walking as a valuable aerobic form of exercise, a 'year-round, readily repeatable, self-reinforcing, habit-forming activity and the main option for increasing physical activity in sedentary populations' (HMSO, 1998: 18).

Although the most fundamental and seemingly 'natural' mode of transport for us bipeds, walking is informed by various performative norms and values which produce distinct praxes and dispositions. Like other forms of travel and tourism, walking can be conceived of as a 'search for a vantage point from which to grasp and understand life' (Adler, 1989: 1375) and to transmit identity. Walking entails 'movement through space in conventionally stylised ways', and is evaluated by

Body & Society © 2000 SAGE Publications (London, Thousand Oaks and New Delhi),
Vol. 6(3–4): 81–106
[1357–034X(200009/12)6:3–4;81–106;015472]

shared tenets of performance, which 'serve as a medium for bestowing meaning on the self and the social, natural or metaphysical realities through which it moves' (Adler, 1989: 1366–8). Accordingly, walking bodies communicate meaning through rhythms and gestures, constituting racial, ethnic, class and subcultural allegiances which are 'signalled, formed and negotiated through bodily movement' (Desmond, 1994: 34). As I will shortly discuss, the rise of excursive walking in the Romantic era is part of the development of modern corporeal reflexivity. Walking in the countryside evolves into a practice designed to achieve a reflexive awareness of the self, and particularly the body and the senses. Nevertheless, like many other everyday physical enactions, walking is often an unreflexive and habitual practice which unintentionally imparts conventions concerning the 'appropriateness' of bodily demeanour, but which is not wholly determined by cultural norms.

The countryside is partly produced by the regular routes which walkers follow. The patterns created by these movements through space are described by David Seamon (1979) as 'place ballets' which delineate a compendium of dances played out in locales. As a geographically and historically located practical knowledge, walking articulates a relationship between pedestrian and place, a relationship which is a complex imbrication of the material organization and shape of the landscape, its symbolic meaning, and the ongoing sensual perception and experience of moving through space. Thus besides (re)producing distinctive forms of embodied practices (and particular bodies) walking also (re)produces and (re)interprets space and place. Besides inscribing paths and signs in rural space, along with specific patterns of erosion, pedestrian bodies also delineate particular kinds of landscape as suitable for particular kinds of walking.

In this article, I will explore ideas and techniques of walking through the British countryside to reveal distinctive ways in which we express ourselves physically, simultaneously performing and transmitting meaning while sensually apprehending 'nature' and sustaining wider ideologies about nature, and the role of the body in nature. First, then, I will identify the Romantic origins of modern walking as a means to inculcate reflexive response to nature and the body, and explore the enduring discursive notions which are embodied in walking praxes. I will move on to consider more broadly the practical conventions which stem from these and inhere in distinct forms of walking and the different ways in which adherents express identity and claim status through celebrating particular physical experiences and by using their bodies to signify and transmit values. I will follow this by examining the disciplinary codes and techniques which organize walking bodies. The analysis is intended to suggest that walking in the country, widely proclaimed as a 'natural' activity which frees the individual and the body from quotidian routine and physical confinement, is beset by conventions about what constitutes 'appropriate' bodily conduct, experience and expression. However, although walking appears to be a far from unmediated pleasure, my final section will raise questions about the disruptive potential of rural walking by exploring the work of Richard Long, bringing back into focus the specificities of rural space and the physical confrontation of the walking body with the contingent, the sensual and the irregular.

Walking as a Reflexive Practice

In medieval times, walking was usually bounded by an individual's 'day's walk circle', an area within which most everyday activities and adventures were confined (Wallace, 1993: 26). To venture beyond these confines was likely to be dangerous, gruelling and viewed as potentially criminal, as the figure of the 'footpad' suggests (Jarvis, 1997: 23). Moreover, there was a dearth of through tracks, due to the locally constituted boundaries of society and space. Bodies moving through space were subject to the regulation of feudal law or, by virtue of their remove from the usual tightly bound social networks, were outside the pale, an exclusion that was all too evidently inscribed on their bodies through poverty, nakedness, insanity and starvation (Urry, 2000: 51).

With the advent of cheap, reliable travel in the 18th and 19th centuries, which opened up a variety of accessible destinations, and a regularized working week with apportioned leisure time, the development of walking as a popular pursuit generated a set of ideological and aesthetic notions. Walking provided a contrast to these new, speedier forms of travel, heightening awareness of the distinct sensibilities and perceptions it facilitated. The art of walking that emerged in the Romantic era promoted a set of interlinked reflexive conventions, aesthetic imperatives and practical endeavours which produced a distinctive relationship between the walking body and nature, cultivating new forms of subjectivity and ideas about nature, and diminishing the association of walking with poverty, criminality and homelessness.

Wallace contends that Wordsworth, a key progenitor of Romantic walking, presented the heroic walker as a metaphorical figure who could resolve the 'aesthetic problems raised by rapid industrialization and the effects of the transport revolution and enclosure' on the individual (1993: 9). Through connecting places and following a 'continuous yet moving perspective', walking could restore the 'natural proportions of our perceptions, reconnecting us with both the physical world and the moral order inherent within it'. At a time of enormous social change, the walker could recover the values of the past (Wallace, 1993: 13), and re-establish continuity. The symbolic figure of the farmer (as cultivator and preserver of the agricultural cycle) in the earlier Georgic rural aesthetic, suffered a loss of stability with the advent of rural change and the new possibilities suggested by speed and technological innovation. Wallace argues that Wordsworth replaced the farmer with the 'excursive walker', for the process of walking entailed 'continued movement, continued process, continued expansion' (1993: 74). Hence the walker is able to resolve transformation by recovering past value, experiencing continuity, embracing change, while acquiring poetic sensibilities.

What I take to be important from Wallace's account is a wider, enduring shift in the relationship between bodies and the countryside whereby the labouring body of the farmer, his identity intimately bound up with his consistent and cyclical toil over a fixed acreage of land, is displaced by the reflexive body of the walker, constantly moving through an aestheticized rural space. The 'body-in-becoming' of the walker, as opposed to the 'body-in-being' of the farmer, is subject to an often intense, reflexive monitoring about the way in which it moves through, senses and apprehends nature – a reflexivity that finds expression in poetry, essays and guidebooks and becomes institutionalized in a range of walking practices. This reflexivity is enmeshed within evolving formations of the modern self, notably in response to the unnerving development of an industrial and urban economy. Thus walking becomes bound up with notions of individuality and self-development, with a retreat from the city and the urban self, and towards a freeing of the body, a rediscovery of childish sensation, and aesthetic and moral regeneration. As Jarvis puts it, Romantic walkers were 'intent on clearing an autonomous space for themselves, in which the self could be reduced, physically and intellectually, nearer to its essentials' (1997: 40). These concerns continue to be espoused by walkers, and are embodied through the distinct practices they enact, highlighting how the modern self as a reflexive project extends to the body and awareness about its sensual experience, appearance, health and performative abilities (Giddens, 1991). I will trace out some of the discursive specificities that construct these notions about reflexivity, as part of what Foucault calls 'a critical ontology of self' (1988), by looking at a range of writings, spanning the early modern period to the present, to identify some of the continuities by which the body is trained to understand, perform and experience walking in the country.

Walking is widely conceived as 'a valuable and enjoyable antidote to the increased uncertainty and tension that unfortunately are so often features of modern life' (Duerden, 1978: 1). Wainwright, doyen of fell-walkers, proposes walking as 'the perfect tonic for a jaded mind and a cure for urban depression' (1969: xix). The urban–rural dichotomy is sustained by ideas about the value of walking in the country, which assert the beauty, freedom, social and natural order of the rural by contrast with the urban. The performances of walkers reinscribe these dichotomies through their reflexive deliberations and spatial practices.

More specifically, the urban is seen as constricting, as the current pamphlet advertising the Ramblers Association states:

> . . . living means being free to roam, to step out, without pointless restrictions, free as the wind, through the woods and over the common. . . . Escape into green country today. And tomorrow. There is something missing in everyday urban life. Out beyond lies the open countryside – a reality of freedom unknown to many.

These widespread anti-urban notions are implicitly echoed by popular ideas and contemporary theories about how bodies are restricted in the city: by strategic surveillance, policing techniques, CCTV and aesthetic monitoring (see Edensor, 2000). Famously, Michel de Certeau describes how walking is tactically used by urban pedestrians to create contingent 'spaces of enunciation' (1984: 98) by composing a path, a fleeting creative inscription which attempts to avoid these constraints. In an earlier age, Simmel (1971) discusses the adoption of a blasé attitude to minimize the assault on the senses in the city, a condition which is compared with a pre-urban realm of relative calm where such defences are unnecessary. In a more recent account, Richard Sennett (1994: 15) argues that urban space has largely become 'a mere function of motion', engendering a 'tactile sterility' where the city environment 'pacifies the body'. Urban movement is apparently typified by rapid, mechanized transit without arousal. The 'micro-movements' used to negotiate space are minimal, producing a desensitized effect, and the 'desire to move freely' and quickly 'has triumphed over the sensory claims of the space through which the body moves'. Similarly, Trevor Boddy (1992) describes a set of 'new urban prosthetics', a comprehensive movement system of smooth and sealed walkways, escalators, bridges, people-conveyors and tunnels which direct movement and carry people, linking them with working, recreational and commercial spaces. Such systems, he argues, anaesthetize bodies, reducing their movements so that they never perform 'a clenched fist, a passionate kiss, a giddy wink, a fixed-shoulder stride' (1992: 123–4). Urban bodies in movement, according to these formulations, are highly regulated, defensive, passive, sensually deprived, performatively inert and, therefore, not conducive to reflexive practices. By contrast, the walking body in the country is conceived as being released from these restrictions.

Primarily, the sensual constraints of the city are imagined to fall away in the country. Romantic walking practices have typically been bound up with the search for the sublime and the picturesque, and as a means to enact the 'romantic gaze' (Urry, 1992). Walking can seem to be a strategy for capturing particularly treasured views unobtainable by other modes of transport. Jarvis demonstrates that the typically non-linear progress of the walker is synonymous with the idealized irregularity of the picturesque as advocated by Gilpin: the walker wanders over the landscape as does the eye as it seeks out picturesque beauty (Jarvis, 1997: 56). Such accounts certainly capture the 'imperialism of the eye' (see Urry, 1992), the pre-eminence accorded to the visual sense in Western culture, and yet reflections about walking have not been dominated by ocular pursuits or impressions (although, as I will show, some forms of walking are shaped around visual practices). Instead, walking in a rural environment is frequently cited as more extensively stimulating sensual excitation, at variance to the dulled senses of the urban body. The body is believed to 'come alive' in the country; in Thoreau's words, 'walking returns the walker to his senses' (Wallace, 1993: 187). The walker is not merely an onlooker, but experiences nature as tactile and taste-full:

> His pores are all open, his circulation is active, his digestion good. . . . He knows the ground is alive; he feels the pulses of the wind and reads the mute language of things. His sympathies are all aroused; his senses are continually reporting messages to his mind. Wind, frost, rain, heat, cold are something to him. He is not merely a spectator of the panorama of nature, but a participator in it. He experiences the country he passes through, – tastes it, feels it, absorbs it . . . (Burroughs, 1875: 37)

The recovery of sensual experience continues to be a common theme in walking literature and a goal sought by walkers. Romantic concepts of walking include the idea that particular aesthetic and mental stimulations are 'inseparable from the physical conditions of movement through space' (Leed, 1991: 72). These sensations can free the mind and generate reflexivity, whether through philosophical and intellectual thinking or aesthetic contemplation, states of mind that are believed to be difficult to achieve in an urban context.

According to Robinson, 'as one enters the variety and movement of the outside world, the space for interior wandering also grows' (Robinson, 1989: 22), a notion which metaphorically aligns freedom of movement *and* thought, thereby 'detaching the individual from their place in the social structure (and loosening) the moorings of their culturally constructed self' (Jarvis, 1997: 37). And yet it is clear that walkers tend to cultivate dispositions and techniques which promote inner reflection:

> I offer myself to unpredictable occurrences and impingements. The world flows past my body, which may block, pleasurably or uncomfortably, some sudden cometary intrusion and create a *situation*. But mostly I can modulate the immediacy of random intrusions for the sake of encouraging, unimpeded, the 'inner life'. (Robinson, 1989: 4)

These reflexive practitioners are searching for self-actualization and self-restoration, through what has been called the 'walking cure', or 'psychotherapeutic walking' (Wallace, 1993: 7). Injured or alienated by modern urban living, the assumption is that we can walk our way into physical and mental condition. As a contemporary commentator remarks, 'walking will do more for the physical and mental well-being of the vast majority of people than any amount of time in front of a television set' (Duerden, 1978: 7).

A key figure in Romantic conceptions about identity is the child, whose access to an unmediated sensuality and unself-consciousness is highly prized. Thus, accompanying therapeutic notions is the Romantic idea that the inauthentic self can be cast off and the 'primitive', more 'natural', childish sensibilities buried under the over-socialized urban individual allowed to emerge. William Hazlitt believes that while walking 'it is yourself, the profound history of your "self" that now as always you encounter . . . long-forgotten things . . . burst upon my eager sight, and I begin to feel, think and be myself again' (in Robinson, 1989: 17). This early Romantic account of the recovery of a sensual, bodily aware 'child within', occasioning the restoration of a perceptual innocence, echoes contemporary popular sentiments from which critical and intellectual spheres are not immune:

> What people in advanced societies lack (and counter-cultural groups appear to seek) is the gentle, unselfconscious involvement with the physical world that prevailed in the past. . . . Nature yields delectable sensations to the child, with his [*sic*] openness of mind, carelessness of person, and lack of concern for the ordinary canons of beauty. An adult must learn to be yielding and careless like a child if he wants to enjoy nature polymorphously. He needs to slip into old clothes so that he could feel free to stretch out on the hay beside the brook and bathe in a meld of physical sensation. (Tuan, 1974: 96)

To further develop this quest for a more natural state of being, rural walking is also understood as an escape from the 'inauthentic' enactions of everyday urban life, moulded by 'over-civilized' norms of behaviour. Csordas describes the body in the city as 'primarily a performing self of appearance, display and impression management' (1994: 2). Whereas, according to Leslie Stephens, while walking in the countryside, 'you have no dignity to support and the dress-coat of conventional life has dropped into oblivion' (in Mitchell, 1979: 18). An earlier writer on walking renders the comparison with the city more explicit, for there 'each feels that the eyes of the world are upon him, and always he is subconsciously occupied in conforming himself to the world' (Haultain, 1915: 217–18).

The pressures of maintaining an urban appearance are accompanied by those producing sensual overload. Echoing Simmel's notion of the blasé attitude, Stephens goes on to describe how 'a London street is full of distractions, they become so multitudinous they neutralize each other. The whirl of conflicting impulses become a continuous current' (in Mitchell, 1979: 35). By contrast, a rural walk is more peaceable, with sensations felt at a slower rhythm. This rural impression contrasts with the urban experience of temporality which thwarts any attempt at composed, reflexive thought, and the body is regulated by the rhythms of industrial imperatives. The temporal discipline of the city, according to Robert Louis Stevenson, constrains the urban inhabitant, who is 'so dominated by clocks and watches and chimes that he has no time to live' (Marples, 1959: 151). In a similar vein, contemporary environmental pressure group Common Ground sloganize: 'Slow down: Wisdom comes through Walking, talking and listening'.

The reflexive constructions of walking identified above prioritize different goals, including heightened sensory awareness, a restoration of 'authentic' being, a higher state of consciousness or intellectual focus. And yet all crucially express the interlinked ideas that such goals can be achieved in the rural realm, that walking is the practice best designed to achieve them and the body is the site for such states to be generated.

Kinds of Walking: Contesting Bodies

Although they continue to dominate the meanings that circulate around contemporary walking practices in the British countryside, the reflexive bodily understandings identified above do not constitute a single practical and ideological thread, for different strands are embodied in the diverse practices which walkers follow. This proliferation of walking practices is referred to by Kay and Moxham, who maintain that 'recreational walking is so diverse and dynamic that it merits careful classification of its many different forms' (1996: 174–5). They contend that rural walking for leisure can be distinguished according to two groups of walking practices. In the first group are 'sauntering', 'ambling', 'strolling', 'plodding', 'promenading', 'wandering' and 'roaming', conventional forms of walking which are easy, casual, capable of spontaneous participation by groups of mixed abilities, relaxing and sociable. In the second category are 'marching', 'trail-walking', 'trekking', 'hiking', 'hill-walking', 'yomping' and 'peak-bagging', esoteric and minority activities which are strenuous, rigorous, challenging and rewarding, and require planning. And in between these two groups come forms of walking such as rambling. No doubt walking practices could be categorized in many different ways, but the account succeeds in pointing out both the varieties of walking and some of their characteristics.

To give some idea of the various and contested modes of walking and the reflexive values they espouse, I will discuss particular practices and the priorities they embrace, and highlight the tensions that surround a number of key issues about walking. I will look at issues about whether to walk alone or accompanied, whether to follow a marked track or map, whether to engage in sustained walking, and the values associated with 'challenge walking'. These contesting practices generate embodied forms of distinction which relate to particular notions of the body and its relationship with nature.

To Walk Alone or Accompanied?

Early Romantic walkers were apt to champion specific personal qualities: detachedness, dynamism, passion and difference from the crowd, that walking both reflected and developed. Indeed, Jarvis makes a plausible case that the first practitioners of walking for pleasure were intent on asserting their individuality and autonomy through walking, partly as a means of rebellion against the bourgeois norms, not least the favoured travel venture, the 'Grand Tour', into which they were inculcated, challenging the disreputable associations that walking kindled among polite society (1997: 28). Many walkers continue to espouse this individualistic orientation although it has been challenged by more collective forms.

Writers wrestle with the dilemma of whether walking alone or accompanied is preferable. Some propose that the walk is a marvellous device for strengthening and developing friendship, being conducive to conversation outwith the bounds of normal intercourse, offering the chance to share impressions and thoughts. Others vouch for the delights of self-development, communing with 'nature' rather than people, the cultivation of self-reliance and contemplation, and the uninterrupted sensual experiences of nature. Their accounts implicitly evaluate and advocate particular bodily performances, presentations and experiences.

Those who advocate solitary walking place it above these communal values. Barron argues that 'there are times when it is good for a man to walk alone; nature has her privacies, and won't reveal them to you nor me when others are listening' (Barron, 1875: 324). In stronger terms, Robert Louis Stevenson asserts that 'there should be no cackle of voices at your elbow, to jar on the meditative silence of the morning' (in Marples, 1959: 151). Such arguments assume that the countryside must be experienced in 'unmediated' fashion if the walker is to discover revelation in nature and the self. Other bodies and the sounds and actions they produce are apt to disturb physical and mental communion with nature. G.M. Trevelyan argues that, 'when you are really *walking*, the presence of a companion ... disturbs the harmony of body, mind and soul ... made one together in mystic harmony with the earth' (in Mitchell, 1979: 61). These practices hint at the development of a refined bodily disposition, a claim that becomes more explicitly status-oriented when solitary walking is more crudely promoted as superior in contrast to collective walking practices.

Disdain for the 'hordes' ranges from mild to hostile. For example, Marples assumes that the great mass of walkers 'probably appreciate few of the finer points of walking. . . . Their pleasures are mainly physical, perhaps (without offence) one might say animal' (1959: 182). In a similar vein, Sidgwick refers to groups who 'stride blindly across country like a herd of animals . . . desecrating the face of nature with sophism and inference and authority, and regurgitated Blue Book'. 'What of their immortal being?' he asks, and avers that

> . . . it has been starved between the blind swing of the legs below and the fruitless flickering of the mind above, instead of receiving, through the agency of a quiet mind and a co-ordinated body, the gentle nutriment which is due. (in Mitchell, 1979: 59–60)

A more recent book warns of the hazards of independent walking and offers guidance on:

> the highly feared DREADED OTHER PEOPLE who are always in your way, going where you want to go. . . . They tend to cluster in large numbers around people who are trying to avoid them. Like stampeding cattle they destroy everything in their path. (Booth, 1996: 1)

A lack of individuality results from a failure to adopt those techniques which engender bodily and mental coordination. It is clear that the solitary, sensitive walker, developing and refining sensual and intellectual capabilities, is contrasted with these animalistic corps, quite literally incapable of anything other than the most basic motor skills, who are reduced to their (collective) body, and are despised for their insensitive intrusion, blundering into nature and into the solitary walker's consciousness.

On the other hand, rambling was established and is organized as a collective walking practice. The pace of the walk is set to optimize group enjoyment, and experience is communally constituted. The walk is an occasion for sociability, where exercise is taken and countryside enjoyed in combination with convivial chatter and companionship. Early working-class walking ventures into the countryside were established in the 1830s to escape from factory towns and cities, less to mobilize a Romantic gaze than 'to regain good fellowship amidst the mountains and dales away from the antagonistic relations of the factory' (Hill, 1980: 15). Rambling retains these values of companionship and shared experiences above the development of individualistic sensibilities, and is a manifestation of a collective corporeality, a body which draws the disdainful comments exemplified above but is considered a source of collective strength by ramblers. This is especially the case where the physical presence of walkers challenges established relationships between country and walkers. The notion that the countryside is the preserve of the 'shooting classes' was famously contested by the largely working-class national Federation of Rambling Clubs formed in 1905 (Bunce, 1994: 117), leading to the formation of the Ramblers Association which led the famous mass trespass on Kinder Scout in 1932. These early ramblers inscribed their political consciousness about land ownership and rights of way by collectively placing their bodies in the land (Holt, 1995).

Trailing Nature and Mapping Space

From about 1800, walking guidebooks were published (Marples, 1959: 78) which recommended and charted particular routes, identified specific sites and viewpoints, and were directive about where and how to walk. The advent of widespread way-marking and the evolution of organized commercial walking tours was disdained as a domestication of nature by certain walkers (Murray, 1939: 294), and the idea of following a path was dismissed as an unsuitable practice. For Wordsworth and others, those spaces imagined as 'untrodden' were conceived as the most sacred and special (Robinson, 1989: 7). A letter from *The Times* captures the attachment to the idea of the unplanned route and the conception of nature that this evokes:

> To steer one's way through the solitudes by following streams or ridges, or by aiming for far-distant landmarks, and even, on occasion, to be lost temporarily in mist, is . . . part of the 'fun of the fair'. Wild nature, tamed and domesticated, is no longer wild nature: 'man meddles only to mar'. (Murray, 1939: 294)

This notion tacitly constructs a relationship between individual and nature that is maximized by the avoidance of all human intervention, particularly the disciplining of bodies and the shaping of the landscape. The quest to get 'off the beaten track' persists (Buzard, 1993).

The sheer range of contemporary paths – which may be themed according to historical and natural interest, level of difficulty or diversity of habitat – means that there has also been a proliferation of walking practices. For instance, particular routes are designated as paths to discover particular botanical or geological specimens, to 'discover' traces of archaeological interest, or to compile ornithological checklists. These sorts of practices appear to utilize walking as a means to collect sights and foreground visuality through the anthropological as opposed to the Romantic gaze (Urry, 1992), and Jarvis considers that such pursuits are typical of the 'categorical ordering of information' (1997: 46). Moreover, walkers may physically follow pre-modern routes such as Roman roads, pilgrimage paths and medieval drove roads. In this way, the British countryside has become intensively mapped by way-markers, anathema to the individualist walkers who seek nature unmediated by dense representation and contextualization.

Two particularly mapped and disciplinary forms of walking may be referred to here. First, there is the nature trail, usually of less than three miles, where walkers are directed to markers along the route which signify points of interest, often described in detail in an accompanying leaflet. Second, there are the long-distance footpaths which are minutely covered in trail guidebooks, both by the reproduction of large-scale maps covering every part of the walk, and detailed directions about how to follow the path. Such guides tend to divide the walk into sections, which suggest that there are logical places to start and end a day's walking, and they encapsulate the section as encompassing particular themes of interest. For example, a guide to the Alfriston to Eastbourne 'section' of the South Downs Way typically advises, 'Cross the road at the bus stop by the entrance to the country park, between a cycle-hire centre and a cottage, to a kissing gate marked with an acorn symbol' (Millmore, 1990: 42). Interspersed with these highly detailed directions, the walker's gaze is specifically drawn to views and features in the landscape. The same section commends the walker to 'turn round at this point to look at a board explaining the local geology and the formation of the cut-off meanders' (Millmore, 1990: 42–3). The particular physical manoeuvres, directions, gazes and other procedures that the walker is called upon to perform seemingly restrict the variety of potential physical options. In this mapped space, walkers will, if they follow the instructions of the text, stop and read at regular intervals to check they are proceeding correctly and taking in the points of interest, simultaneously travelling virtually, through the map, and actually, through the landscape. These walking practices seem akin to the highly coordinated, collective movements of package tourists which are saturated by interpretation, directedness, procedure and segmentation (Edensor, 1998). Such walking modes fashion a regular and routinized choreography, imposing a rather rigid 'place ballet' (Seamon, 1979: 58–9) on the landscape and a set of practices upon walkers' bodies. Once established, such routes provide a de facto framework for walkers that makes other pathways invisible.

The mapping of walks upon space constructs the notion that there are apposite ways to make paths. Trevelyan recommends that 'road and track, field and wood, mountain, hill and plain should follow each other in shifting vision' (in Mitchell, 1979: 70). Others advocate particular themes and types of scenery. Yet, as Wainwright elucidates, it is commonly understood that there are certain rural spaces in which walking is not fruitful:

> Forest walking is the antithesis of fell-walking, for in the one there is a severe confinement, a rigid line of march, a lack of living creatures, absence of birdsong, inability to see ahead or look around; but in the other is freedom, freedom to roam and explore, to look into far distances. . . . The forest is a prison, the fell is liberty. The one is artificial, as man made it, the other natural as God made it. (Wainwright, 1969: 33)

Thus the propensity to map space for walking is designed to maximize particularly valorized rural environments as opposed to others, particularly those at the margins of rurality such as wasteland, large-scale intensively farmed land and semi-suburban terrain. And maps and guides contain elaborate guidance about how bodies should move and conduct themselves. Most obviously, path-making habitually normalizes a walking practice that emphasizes linear progress through space.

How Far and Fast? Long-Distance and Challenge Walking

The distance of a walk is another factor which implicates notions about the walking body, its practices and proclamations. Some Romantic walkers were famous for the prodigious distances they covered and, since then, long-distance walking has been popularized, carrying with it specific notions of achievement which may serve as a notable chapter in self-narratives. For instance, Wainwright affirms that completing the Pennine Way 'offers you the experience of a lifetime' but certainly not 'continuous entertainment'. Instead it is a:

> ... tough, bruising walk and the compensations are few. You do it because you want to prove to yourself that you are man enough to do it. You do it to get it off your conscience. You do it because you count it a personal achievement. Which it is, precisely. (Wainwright, 1969: xiii)

These idealized spartan endeavours are evaluated as producing a superior physical condition and more intense bodily experiences to the over-socialized, pampered, slothful bodies of everyday life. The construction of pleasure here relies upon the idea that walking of this type forms character through masculine fulfilment. Wainwright concedes, 'make no mistake: you are going to suffer, you are going to get wet through, you are going to feel miserable and wish you had never heard of the Pennine Way', yet he cautions, 'if you start, don't give up, or you will be giving up at difficulties all your life' (Wainwright, 1969: 170). The challenge is to overcome the physical privations and discomforts such a walk promises in order to test one's character: the battle is against nature and against the over-socialized self. Beyond the trial of physical endurance and mental strength lies the promise of a more confident self and a return to a masculine (bodily) essence, replete with fantasies about getting back in touch with (one's) nature. In a more radical form of long-distance walking, those who have completed the 2,500-mile Appalachian trail not only must overcome physical and mental challenge but also the real dangers of wildlife, disease, hypothermia, assault by humans and severe weather. The perils of 'wild' country have been overcome, marking the walker as a pioneer who has completed a significant test (Luxenberg, 1996: 97–103).

The goal-driven pursuit of long-distance walking appears to advance a differently aestheticized body to that of those who promote walking as facilitating a sensual immersion in nature. Although I have shown above that a fit body is esteemed precisely because it enhances the sensitive appreciation of nature, certain accounts of physical prowess veer towards a more austere bodily aesthetic. David Matless (1995) has shown how the inter-war ramblers were concerned to pursue moral and physical achievement through the 'art of right living', a set of a working-class, leftist concerns which extol the virtues of spartan discipline and the pleasures of hard physical exercise. Ramblers often distinguished themselves from other walkers by the distances they covered and the speed at which they walked (Holt, 1995: 13). The ascetic conditions of Youth Hostels and the value of eating wholesome food are set against the indulgent and Epicurean habits of the bourgeoisie (Holt, 1995). Similarly, the tautening of the body's muscles and the happy confrontation with rough weather, privation and hard exercise were contrasted with the pampered bodies of the middle class. This aesthetic adulation of the naturally conditioned body finds an echo in the 'Strength through Joy' movement in Germany, which some British walkers enthusiastically researched (Holt, 1995), but it also has earlier roots. For instance, Barron, in his book, *Footnotes, or Walking as a Fine Art*, declares that a 'muscular, manly leg, one untarnished by sloth or sensuality, is a wonderful thing' (Barron, 1875: 13).

Although most notions of walking foreground the male body, it is particularly prominent here. In the 19th century, its advocates assumed that walking was a male preoccupation. The supposedly delicate bodies of (middle-class) women were deemed unsuited to the sturdy demands of walking. Accordingly, where women did walk, male escorts accompanied them on their travels. Despite this, some women resisted the patronizing constraints that these masculinist practices and beliefs generated (Marples, 1959). Although such ideas may seem outdated, a masculinist aesthetic persists, partly drawing on archaic notions about male exploration and conquest of a passive, feminized nature (Jarvis, 1997: 58).

While the long-distance forms of walking identified above are primarily about self-development and claiming status through bodily achievement, walking as a competitive sport is almost bereft of any Romantic view of the countryside and the cerebral pleasures celebrated by other walkers. Rather, the sporting prowess of the body eliminates other values. The origins of competitive walking stem from the 18th-century exploits of Foster Powell and Captain Barclay, who gained widespread fame and attracted a gambling industry by completing many prodigious feats of 'toe-to-heel' walking, covering huge distances in specific periods of time (Marples, 1959; Murray, 1939). This specialized competitive pursuit later extended to peak-baggers in the 19th century and working-class Manchester pedestrian clubs (Marples, 1959). The dominant contemporary form of sporting rural walking is 'challenge walking', where the goal is 'to overcome the personal challenge of completing a route, usually within a fixed time scale' (Castle, 1991: 1). Challenge walks are organized primarily by the Long Distance Walkers Association and the British Walking Federation. The best-known, the 40-mile Lyke Wake Walk, which must be completed in 24 hours, is undertaken by about 8000 walkers each year. Other events include the 'Fellsman Hike', a 59-mile trek across the Yorkshire Dales, the 'High Peak Marathon', a 4-member team event which must be completed in 18 hours, and the 'Lake District Four 3000s' which involves the ascent of all four Lakeland peaks over 3000 feet within 22 hours. The object is not only to win but to excel and improve upon previous individual performances.

In long-distance and competitive walking, like adventure sports, fulfilment takes the rhetoric of individual achievement counterposed to the regulations and fetters of everyday family and work life, 'connected to a "can do" philosophy of personal growth and reflection' (Cloke and Perkins, 1998: 208). This achievement is accompanied by the sensual experience of the sporting body, with 'heightened sensory experience, risk, vulnerability, passion, pleasure, mastery and/or failure' (Cloke and Perkins, 1998: 214). The finely tuned sporting body, a product of training regimes, is monitored to ensure maximum efficiency as it moves through the country. However, these competitive, physical desires are scorned by many walkers. For instance, although Wainwright values the achievement gained by walking the Pennine Way, he draws the line at challenge walks:

> Inevitably, but mistakenly, the Pennine Way will be subject to record-breaking. Somebody, someday, will write to the papers to proclaim that he has walked the distance in 10 days. Then somebody else will better that, and so on. In due course, the record will be reckoned in split seconds. All this is *wrong*. The Pennine Way was never intended to be a race against time. No, to be enjoyed (and why else do it?) it should be done leisurely. There is much to explore, much to observe, much to learn. (Wainwright, 1969: ix)

In a wider context, by espousing particular cultural values through their embodied practices, walkers acquire status among like-minded practitioners and assert these values to claim that their particular practices are 'appropriate' in a rural setting. This becomes more crucial as arguments multiply about how the countryside should be used; as what sort of 'playground' (Bunce, 1994: 111–40). Whereas walking, along with visiting historic sites, nature study, picnicking, sightseeing and fishing are conceived of as traditional rural leisure activities, they are being supplemented by growing popularity of other ways of moving through the countryside: snow skiing, mountain biking, driving all-terrain vehicles, orienteering, windsurfing, white-water rafting, paint-ball and hang-gliding. These more vigorous and sporting physical pursuits are typically individualistic, competitive, high cost and mechanized, in contrast to the more sedate and cheaper traditional pastimes (Butler, 1998: 212–16). Where walkers have carved out access routes, these may be also used by mountain bike enthusiasts, and a solitary walk to the top of a peak may be accompanied by the adrenaline charged leaps of hang-gliders. In certain areas, such as the Scottish Cairngorms, battles between walkers, skiers and other adventure sport enthusiasts have been played out. While walkers generally want to roam through 'unspoilt areas', the new adventure sports add new symbolic meanings and place-myths to rural settings which emerge out of the embodied practices of enthusiasts (Cloke and Perkins, 1998: 186). Thus the rural becomes imagined and used as a stage for different sorts of exciting physical adventure.

Learning to Walk: Discipline and Walking Techniques

Although many of the walking praxes identified above stress the freeing of the body during a walk in the countryside, their proponents frequently advocate a set of bodily techniques, forms of physical training and valued equipment through which walking can be accomplished. This evokes Mauss's notion of bodily techniques which are technically constituted by a specific set of movements, acquired through training and functional to a specific aim – in this case, walking in the countryside (Williams and Bendelow, 1998: 50). Thus while the affective values, modes of experiencing rurality and bodily practices that constitute distinct walking dispositions are often conceived as an escape from the regulatory framework of everyday urban life, they are governed by 'expert' knowledge and body discipline. The reflexive monitoring and control of the walker's body seemingly contradict these desires for sensual immersion and physical and mental freedom. But also, 'bodily hexis', a specifically *unreflexive* mode of moving and feeling through which 'physical capital' is transmitted, is the embodiment of the walker's habitus (Bourdieu, 1977).

Among serious walkers, a host of conventions maximize comfort and safety. The walking body needs to be trained and conditioned to attain the level of fitness required for strenuous hill-walking. Moreover, the much-vaunted sensual experience of walking is partly dependent upon the fitness of the walker's body. The body in nature is conceived as healthier and fitter and thus more able to sense and to feel, to be more aware of itself and its 'natural' propensities, as G.M. Trevelyan, a popularizer of walking, describes: 'the body, drugged with sheer health, is felt only as a part of the physical nature that surrounds it and to which it is indeed akin' (Mitchell, 1979: 76). In addition, dietary provision is deemed an essential part of this bodily management. To ensure that the walker does not run out of energy, an awareness of food intake means that the walker monitors the body by eating 'little but often' (Duerden, 1978).

Other institutionalized codes ensure a more collective form of bodily discipline. For instance, group walking is often a hierarchical affair, organized by a 'leader' who, according to a recent manual, 'should be a person well experienced in mountain walking, energetic, determined, sympathetic, cheerful yet cool in adverse situations' and should also 'aim to obtain maximum interest from the scenery and the environment' (Williams, 1979: 76–7). The leader is advised to choose the distance and route, and devise progressively more difficult expeditions for novices and groups should 'follow in crocodile style behind the leader [keeping] a few paces between each other to observe the ground immediately ahead and must not become strung out' (Williams, 1979: 87).

Furthermore, the emphasis in some guides to walking on map and compass technique requires that walkers continually monitor their direction, seeking out which landmarks and contours to follow. Some writers urge walkers to follow procedures which join sites together by means of 'walking on a bearing', and 'navigating around obstacles', a practice which necessitates a continual awareness of the body and its relationship to space. Walkers are also often trained in safety procedures in the event of accidents and bad weather, and are taught manoeuvres over rock and water. Rather than an uninterrupted occasion for contemplation and sensual pleasure, such disciplines lead to continual physical self-control and spatial orientation.

It might be imagined that walking simply involves putting one foot in front of the other, yet some authorities maintain that it requires particular techniques. Williams recommends that the trekker should 'acquire an easy, effortless walk'. Moreover:

> The body should lean slightly forward to offset the weight of the rucksack. There is little movement of the arms and the hands are kept free. The legs are allowed to swing forward in a comfortable stride. High knee movements and over-striding are to be avoided as they are very fatiguing . . . the pace should be steady and rhythmical and the feet placed down with a deliberate step. As each stride is made the whole of the foot comes into contact with the ground, rolling from the heel to the sole. (Williams, 1979: 94)

Duerden advises that the walker ought to:

> . . . achieve a steady pace with rhythmic strides. . . . The weight of the body should be moved slightly forward, i.e. a slight forward stoop, with a short, smooth swing of the arms. The sure sign of a good walker is the manner in which he makes it all look very easy, as if he could go on all day without tiring.

He also suggests techniques for walking downhill and uphill, cautioning against the 'temptation to rush the last few yards' as false crests are frequent and energy should be conserved (Duerden, 1978: 12). The implication of this advice is that the body must be schooled in correct walking technique and the walker must persistently check the terrain for the requisite performance.

Besides these mechanical approaches to walking, methods for achieving mental stimulation via walking are proffered. Here, I want to suggest that some of the dispositions identified above which regard walking as a means to escape an over-socialized, desensitized self are ironically acquired through reflexive bodily management. For example, Barron celebrates the stimulative value of walking, particularly isolating the legs:

> These members when in motion, are so stimulating of thought and mind, they almost deserve to be called the reflective organs. As in the night an iron-shod horse stumbling along a stony road kicks out sparks, so let a man take to his legs and soon his brain will begin to grow luminous and sparkle. (Barron, 1875: 15–16)

Yet he warns that walking may degenerate into a 'brutish affair' if you 'follow too rapid a gait' since the 'attention may be dissipated'. On the other hand, he asserts that 'a pace too slow begets sluggishness of mind and at once makes an end of all your fine susceptibilities' (Barron, 1875: 10). Other writers focus more particularly on this training of the mind to develop the self:

> The frame of mind [with] which one ought to set out upon a rural peregrination should be one of absolute mental vacuity . . . one ought to rid oneself (of) the categories of time and space. . . . The proper frame of mind is that of absolute and secure passivity; an openness to impressions; a giving up of ourselves to the great and guiding influences of benignant nature; a wondering and childlike eagerness – not a restless and too inquisitive eagerness. (Haultain, 1915: 5–6)

And some authorities specify the kind of terrain that must be walked through to generate this mental stimulation: 'uneven walking is not so agreeable to the body, and it distracts and irritates the mind' (Robinson, 1989: 18). G.M. Trevelyan claims that 'the perfection of walking . . . requires longer time, more perfect training . . . a different kind of scenery than "ordinary" walking' (in Mitchell, 1979: 59–60). These technical notions, which propose that particular forms of walking can be learnt and developed, stem from and engender appropriate ways of interacting with the countryside.

The bodily dispositions inculcated by these techniques and disciplinary procedures are intended to communicate particular values and status positions, and so are the ways in which the walking body is clothed and housed. And the stylistic identities of walkers are well catered for by the market in walking accoutrements, products which range from tents and sleeping bags to clothes, and extend to a host of accessories. Different forms of cultural capital are transmitted via these products. Status-conscious decisions are made by seasoned walkers when buying gear, about the credibility of clothing and equipment, especially around the quality of 'heavy-duty' as opposed to 'fashionable', ironically a status which is gained by reinforcing the notion that walkers do not concern themselves with 'trivial' fashions or self-adornment. Yet a glance at the range of products on display at hiking retail outlets suggests that fashion and style are also important.

Nowhere is clothing more fetishized than in the case of boots, often accorded magical properties in their ability to assist the walker to cover more ground, a fetishization that is accompanied by ritualistic dubbining. This fashion awareness among walkers clashes with notions that walking is conducive to a disregard for the over-socialized, performative self which is resisted by an apparent disregard for appearance, itself, a 'fashion statement'.

The backpack is also a particularly symbolic item that inscribes the walker's body with a badge of membership:

> The heavy-duty backpack, as we now see it, is a symbol of self-sufficiency, and its bearer is proclaiming a message just as explicit as when clothes told of class distinctions. He is saying that he can travel without the help of anyone . . . he is fit and . . . is prepared for any eventuality. The backpack is a badge of intent, a membership card to a confraternity. (Jebb, 1996: 74)

Yet backpacks have been tailored for a variety of walking practices and identities. Scientific advice about particular models warns walkers against being dazzled by style, for backpacks are apt to influence the style of walking. Jebb warns against top-heavy models which cause one to tilt forward so that 'the pure pleasure of walking is lessened' since it 'derives from the co-ordinated rhythm of the upright body, unencumbered by dead-weight of burdens, free-flowing and perfectly balanced' (Jebb, 1996: 75).

Clothing for walking must take account of ease of movement and climate, and so the body must be protected and continually monitored when faced with bad weather. The manufacture of a wide range of cagoules, breeches, socks, over-trousers, shorts, sweaters, padded jackets, gaiters and the like testifies to the demand for a comfortable and 'appropriately' clothed body. These walking accessories can be considered as 'actants' (Urry, 2000: 79), objects which facilitate and constitute the aims and practices of walkers. The distinct walking practices identified above incorporate particularly valued objects which are utilized to achieve particular goals. Thus, the network of specific things employed by challenge, long-distance and casual walkers help to constitute the ideas and values they espouse. Moreover, the hybrid forms combining bodies and things also produce particular sensual experiences and species of movement. Where these human–object networks endure, as they seem to among walkers, they provide a regime through which conventions are maintained, orthodoxies which are given a boost by the niche markets which target particular groups.

Walking Away from Conventions: Sensuality, Disruption, Difference

I have highlighted a variety of walking praxes through the countryside, identifying an acutely reflexive range of often contested practices, which mobilize diverse discourses and techniques and project distinct forms of status through bodily hexis and disposition. This suggests that the walking body, rather than being the site of liberation from sensual deprivation, over-socialization, external and internal surveillance, anxieties about status and inauthentic performance, actually duplicates all these constraints. A reflexive body ties itself into a series of institutionalized cultural resources – discourses, networks, techniques and hybrids – that themselves impose normative codes and controls upon walkers and their bodies. What was mobilized as a reflexive struggle against modern convention and subjection, has instituted its own normalizing, unreflexive codes.

However, lest we forget, the body is the means through which we experience and feel the world; the senses act to inform presence and engagement to constitute a 'being-in-the-world' (Csordas, 1994: 10). Accordingly, cultural meanings and social relations are not only inscribed upon the body, but are produced by it, and the senses both experience and structure space. Bodies belong to places and help to constitute them whether they stay in place, move through place or move towards other spaces (Casey, 1996). Whatever normative practical conventions of walking may prevail, the body can never be assumed to passively perform a dance of duty, for bodies are not only written upon but also write their own meanings and feelings upon space in a process of continual remaking. And, as Jarvis contends, since walkers are 'more alert to the multiplicities and the particularity of actual landscapes . . . [walking] is capable of fostering resistance to any idealizing tendencies' of nature (1997: 69). Moreover, alternative forms of walking challenge normative modes, and individuals may escape these conventions to produce meanings and practices of their own.

Theories about the relationship between the body and (particular kinds of) space ought to neglect neither the material character of space nor the sensual propensities of the body. In other words, the material, spatial, sensual and temporal contingencies of any walk mean that 'the walker is in experience, feels and thinks in his movements through space and time' (Robinson, 1989: 4). For instance, as Game argues, the walking body is moved by affect and its movement 'invokes memories which are involuntary' (Game, 1991: 152)

In the countryside, walking is characterized by distinct material, temporal and sensual characteristics. Wanderers and strollers are likely to be confronted by the contingent, the unfamiliar and the unforeseeable. And like the early modern urban dweller, the *flâneur*, such walkers are less confined by temporal and spatial constraints about where and when they should walk. Moreover, the different distribution of sensory stimuli – the smells, the sounds, the sights, the feelings and the tastes of the countryside – are also part of the ever changing panoply of experience which walking produces. The sensual experiences of a walk in the country may be lingered over due to the pace of travel and the relatively slow speed of things moving past. The events and things that one encounters are conceived by Trevelyan as the contingencies 'and a thousand other blessed chances of the day' which are 'the heart of walking' (in Mitchell, 1979: 69). Thus the country is not more 'natural' than the city but can be roughly characterized by distinct forms of spatial arrangement, a set of differences to which the body is likely to react in particular – though not necessarily predictable – ways.

In walking of all kinds, the body can never mechanically pass seamlessly through rural space informed by discursive norms and practical techniques. The interruptions of stomach cramps and hunger, headaches, blisters, ankle strains, limbs that 'go to sleep', muscle fatigue, mosquito bites and a host of other bodily sensations may foreground an overwhelming awareness of the body that can dominate consciousness. Moreover, the terrain and climate are apt to impose themselves upon the body, irrespective of discourses about the rural idyll and the Romantic countryside. The body must perform certain tasks, which may be painful or pleasurable in their novelty, or challenging in their awkwardness. Walkers must avoid barbed wire, be wary when passing through fields of bullocks, make sure they do not step in cowpats or mud or in holes, step over logs, leap across streams, negotiate stepping stones and stiles, swat swarms of flies away, avoid brambles, nettles and thistles. These actions dramatically involve bodily actions and reveal physical properties. For instance, climbing over an unstable and swaying fence, the walker may become suddenly aware of the body's mass and weight. Environment and climate thus impose upon walking strategies and sensations. The tactile qualities of many rural paths produce a mindfulness about one's balance as well as a practical and aesthetic awareness of textures underfoot and all around. The walking body treads across rocky ground, springy forest floor, marsh and bog, rough tracks, heathery moorland, long grass, mud, root-lined surfaces, pasture, tarmac and autumnal leafy carpets. Biting insects inhabit long grasses, rain drenches clothes, frosty air freezes body parts.

While walking through the countryside can disrupt ordinary urban walking habits because of these sensory, material and imaginative intrusions, Game avers that certain forms of walking can disrupt order better than others. 'Wandering' and 'strolling', for example, are forms of walking which 'err from the straight and narrow of linearity, or the order', and can be distinguished from walks which seek more purposive ends (1991: 150). Likewise, some rural areas invite an improvised and undelineated walking performance whereas others restrict options because the landscape has been shaped to circumscribe well-marked footpaths so as to prohibit or impede access.

Another feature which may interrupt the quotidian is the temporal structure of extensive walks which contrasts with the segmented, timetabled structure of ordinary diurnal patterns. Jarvis considers that walking is more likely to produce a 'progressional ordering of reality' where impressions and sensations engendered by prolonged motion generate a sequential experience (1997: 46). Especially if undertaken alone, extensive walking can produce an acute awareness of small changes in mood, along with the ever-changing landscape. An extended period of walking, perhaps for 12 hours, is a long time to be alone without engaging in purposive activity or being distracted by entertainments and demands. This different temporal structure means that time can be difficult to comprehend. Periods of walking seem much longer or shorter than they are, and other influences produce a temporal pattern – when to rest, to eat – which is shaped more by physical contingencies, the rhythm set by the legs and the nature of the terrain than the clock.

I am suggesting that walking in the country possesses the potential for disruption. According to Richard Sennett, the 'body comes to life when coping with difficulty', is roused by the resistance which it experiences. Moments of confrontation, of self-displacement, are vital to preserve openness to stimuli, to awaken the senses, and an acceptance of 'impurity, difficulty, and obstruction' is 'part of the very experience of liberty' (1994: 309–10). Thus, walking can indeed be particularly suitable for stimulating reflexivity, yet the moment it becomes devised and practised as such, an awareness of practical conventions can obscure the chance occurrences and multiple sensations that stimulate a different sort of reflexivity, one which embraces the difference, the alterity of nature, the contingent, the heterogeneous, the decentred, the fleeting and the unrepresentable. Generated by the walking body, this reflexivity, in contradistinction to the cognitive, instrumental, monitoring reflexivity described by Giddens (1991), promotes a rejection of the practical conventions and codes identified above, and welcomes the unknowability of a heterotopic nature, a sensual body and a multiple sense of place.

To illustrate the points raised above, I want to look at the work of sculptor Richard Long, who is intimately concerned with the experience of walking. His work captures the sensual, material, contingent process of walking and foregrounds physical agency by featuring the traces of bodies, also questioning conventional modes of understanding and practising walking.

Long's choice of route often veers away from the normative selection of well-trodden paths. For instance, in *A Walk by All Roads and Lanes Touching or Crossing an Imaginary Circle* (1977), an Ordnance Survey map of Somerset describes in black ink an arbitrary circle which follows the route walked by the artist, and *A Six-Day Walk Over All Roads, Lanes and Double-Tracks Inside a Six-Mile-Wide Circle Centred on the Giant of Cerne Abbas* (1975) takes this process further by filling the circle in, so to speak, by covering all paths within its circumference. These works are a rebuke to the normative, linear determination of much walking, and yet they also smack of the impossible desire to fully know the spaces moved through. There is also the desire to measure, to quantify, in Long's work, and yet the measurements he employs are far removed from those used in guidebooks. In *From Tree to Tree: A Walk in Avon, England* (1986), he lists the random occurrence of trees at points along the walk, as in:

palm
oak 9 miles
scots pine 15 miles
hawthorn 22 miles

Similarly, in *The Wet Road: The Times of Walking on the Rain-Wetted Road Along a 19-Day Walk of 591 Miles from the North Coast to the South Coast of France* (1990), Long simply cites a sequence which commences 'FIRST DAY $1^{1}/_{2}$ HOURS, THIRD DAY $2^{1}/_{4}$ HOURS' and yet this work strongly suggests the sensual qualities of wet tarmac by isolating this phenomenon in walking time and space, making us conscious of the characteristics of the ground upon which we walk. These modes of marking create a narrative structure which speaks of the need to chart progress, signify process, yet also reveals the arbitrariness of conventional measuring techniques.

Long is particularly interested in tracing the body's path through nature, noting the actions it takes and also the traces it makes. In a work entitled *A Moved Line in Japan* (1983), Long describes a sequence where he picks up, carries and places 'one thing next to another' during a 35-mile walk at the edge of the Pacific Ocean. One section in the 26-line sequence reads:

seaweed to pebble
pebble to dog's skeleton
dog's skeleton to stick
stick to mermaid's purse
mermaid's purse to bamboo

Besides the uncanny poetic beauty of this list, the line is effective in revealing a focusing in of human perception on particular objects and the subsequent actions that involve their removal, and although such traces may never be evident, these changes in the scene identify a body which acts within, improvises, and is part of nature. Traces of the body are more overt in several of Long's other works, such as *A Line Made by Walking*, which marks the traces of movement by the absent human agent. The figure of the path is a key metaphor for Long, who reveals that it is an outcome, a sign of the actions of one or more bodies.

Figure 1 A sixty-minute circle walk on Dartmoor 1984 by Richard Long

Several works bring together the notion that walking is a way of being in the world that combines an experience of the sensual, the serendipitous and the irruptive body during a passage among a material nature. The sensual body is addressed in *Dartmoor Wind Circle: A Walk Eight Miles Wide* (1988) where, during a circular walk, Long charts the direction of the wind at regular intervals (yet another way of measuring), and the representation of the walk takes the form of a series of arrows, aligned with the walk's circular shape, which indicate the wind direction. Minimalist though it is, this work conjures up the sensations felt by the absent body, buffeted by winds from changing directions, and denotes the moving body as one element in a pattern of intersecting movements and energies, impacted upon and impacting upon the world. Likewise, *Sound Line: A Walk of 622 Miles in 21 Days from the North Coast to the South Coast of Spain* (1990) foregrounds the sense of hearing, suggesting sounds arbitrarily experienced and noises created by the walking body, and also highlights the beauty of the sounds of words, as in place-names. In one sequence, A BRAYING DONKEY NEAR SEGURILLA, is followed by KICKING A STONE IN ALCAUDETE DE LA JARA, which is followed by HISSING WIND THROUGH BRANCHES IN

LA NAVA DE RICOMALILLO. Places along the walk thus become distinguished by their sounds.

In a stunningly eloquent and evocative work, *One Hour: A Sixty-Minute Circle Walk on Dartmoor* (1984) (Figure 1), Long brings these themes together, citing the sensations felt and experienced by the body, its movements, the noises and other impacts it makes on nature, the texture and shape of the terrain, the serendipity of things stumbled upon, heard and sighted.

References

Adler, J. (1989) 'Travel as Performed Art', *American Journal of Sociology* 94: 1366–91.

Barron, A. (1875) *Footnotes, or Walking as a Fine Art*. Connecticut: Wallingford Printing Company.

Boddy, T. (1992) 'Underground and Overhead: Building the Analogous City', in M. Sorkin (ed.) *Variations on a Theme Park*. New York: Hill and Wang.

Booth, F. (1996) *The Independent Walker's Guide to Great Britain*. Gloucester: Windrush Press.

Bourdieu, P. (1977) *Outline of the Theory of Practice*. Cambridge: Cambridge University Press

Bunce, M. (1994) *The Countryside Ideal: Anglo-American Images of Landscape*. London: Routledge.

Burroughs, J. (1875) 'The Exhilarations of the Road', in *The Writings of John Burroughs, vol. 2*. New York: Houghton Mifflin.

Butler, R. (1998) 'Rural Recreation and Tourism', in B. Ilbery (ed.) *The Geography of Rural Change*. London: Longman.

Buzard, J. (1993) *The Beaten Track: European Tourism, Literature and the Ways to Culture 1800–1914*. Oxford: Clarendon Press.

Casey, E. (1996) 'How to Get from Space to Place in a Fairly Short Stretch of Time: Phenomenological Prologema', in S. Feld and K. Basso (eds) *Senses of Place*. Santa Fe, NM: School of American Research Press.

Castle, A. (1991) *Challenge Walking*. London: A. and C. Black.

Cloke, P. and P. Perkins (1998) '"Cracking the Canyon with the Awesome Foursome": Representations of Adventure Tourism in New Zealand', *Environment and Planning D: Society and Space* 16: 185–218.

Csordas, T. (ed.) (1994) *Embodiment and Experience*. Cambridge: Cambridge University Press.

de Certeau, M. (1984) *The Practice of Everyday Life*. Berkeley: University of California Press.

Desmond, J. (1994) 'Embodying Difference: Issues in Dance and Cultural Studies', *Cultural Critique* Winter: 33–63.

Duerden, F. (1978) *Rambling Complete*. London: Kaye and Ward.

Edensor, T. (1998) *Tourists at the Taj*. London: Routledge.

Edensor, T. (2000) 'Moving Through the City', in D. Bell and A. Haddour (eds) *City Visions*. London: Longman.

Foucault, M. (1988) 'Technologies of the Self', in L. Martin, H. Gutman and P. Hutton (eds) *Technologies of the Self: A Seminar with Michel Foucault*. London: Tavistock.

Game, A. (1991) *Undoing the Social: Towards a Deconstructive Sociology*. Milton Keynes: Open University Press.

Giddens, A. (1991) *Modernity and Self-Identity*. Cambridge: Polity.

Haultain, A. (1915) *Of Walks and Walking Tours: An Attempt to Find a Philosophy and a Creed*. London: T. Werner Laurie.

Hill, H. (1980) *Freedom to Roam*. Ashbourne: Moorland Publishing.

HMSO (1998) *Walking in Great Britain: Transport Statistics Report*. London: HMSO.

Holt, A. (1995) *The Origins and Early Days of the Ramblers Association*. London: The Ramblers Association.
Jarvis, R. (1997) *Romantic Writing and Pedestrian Travel*. Basingstoke: Macmillan.
Jebb, M. (1996) 'Backpackers', in D. Emblidge (ed.) *The Appalachian Trail Reader*. New York: Oxford University Press.
Kay, G. and N. Moxham (1996) 'Paths for Whom? Countryside Access for Recreational Walking', *Leisure Studies* 15: 171–83.
Leed, E. (1991) *The Mind of the Traveller: From Gilgamesh to Global Tourism*. New York: Basic Books.
Luxenberg, L. (1996) 'Difficulties and Dangers Along the Trail', in D. Emblidge (ed.) *The Appalachian Trail Reader*. New York: Oxford University Press.
Marples, M. (1959) *Shanks's Pony: A Study of Walking*. London: J.M. Dent and Sons.
Matless, D. (1995) '"The Art of Right Living": Landscape and Citizenship,1918–39', in S. Pile and N. Thrift (eds) *Mapping the Subject*. London: Routledge.
Millmore, P. (1990) *National Trail Guide: South Downs Way*. London: Aurum Press.
Mitchell, E. (ed.) (1979) *The Pleasures of Walking*. Bourne End, Bucks: Spurbooks.
Murray, G. (1939) *The Gentle Art of Walking*. London: Blackie.
Robinson, J. (1989) *The Walk: Notes on a Romantic Image*. Norman: University of Oklahoma Press.
Seamon, D. (1979) *A Geography of the Lifeworld*. London: Croom Helm.
Sennett, R. (1994) *Flesh and Stone*. London: Faber.
Simmel, G. (1971) 'The Metropolis and Mental Life', in *Individuality and Social Forms*. Chicago, IL: University of Chicago Press.
South Bank Centre (1991) *Richard Long: Walking in Circles*. London: Thames and Hudson.
Tuan, Y.-F. (1974) *Topophilia: A Study of Environmental Perception, Attitudes and Values*. Englewood Cliffs, NJ: Prentice-Hall.
Urry, J. (1992) 'The Tourist Gaze "Revisited"', *American Behavioural Scientist* 36: 172–86.
Urry, J. (2000) *Sociology without Societies*. London: Routledge.
Wainwright, A. (1969) *Pennine Way Companion*. Kendal: Westmorland Gazette.
Wallace, A. (1993) *Walking, Literature and English Culture*. Oxford: Clarendon Press.
Williams, P. (1979) *Hill Walking*. London: Pelham.
Williams, S. and G. Bendelow (1998) *The Lived Body: Sociological Themes, Embodied Issues*. London: Routledge.

Tim Edensor teaches Cultural Studies at Staffordshire University. He is the author of *Tourists at the Taj* and has written on the film *Braveheart*, Scottish heritage and tourism. He has recently edited and contributed to a book: *Reclaiming the Potteries: Leisure, Space and Identity in Stoke-on-Trent*, and is currently writing a volume about national identity and popular culture.

These Boots Are Made for Walking . . . : Mundane Technology, the Body and Human–Environment Relations

MIKE MICHAEL

It is of no great insight to note that recent sociological thought has increasingly become interested in the role of the body and embodiment in social processes (e.g. Shilling, 1993; Synnott, 1993; Turner, 1992, 1996). One dimension of this 'turn to the body' is a conceptualization that, in Rose's terms, speaks not of the body per se, but of 'how a particular body-regime has been produced, the channelling of processes, organs, flows, connections, the alignment of one aspect with another' (1996: 184). A key intermediary in this process of alignment is technology. Needless to say, the role of technology has been complex and variegated: it can shape our local comportment (e.g. Latour, 1992; Michael, in press); it can structure our conceptions of the body (Birke, 1999; Martin, 1989); it can be paramount in a general ethos of disembodiment (Virilio, 1995; though see Morse, 1994; Slater, 1997). Accordingly, to explore the role of the body in the mediation of relations between humans and the natural environment is, inevitably, to consider the part played by technology. Technology has, as we shall see, shaped our relations to nature in many ways, not least in terms of the means by which we access nature, both physically and representationally (though, of course, these are hardly distinct). Thus, as Macnaghten and Urry (1998), among others, have documented, certain 'epochal' technologies have been crucial in the shifts in our apprehensions of nature – for example, technologies of representation such as photography, and technologies of transportation such as the train and the car. In contrast to these treatments, the present article is concerned not with 'epochal' technologies, but mundane ones, specifically, walking boots.

Body & Society © 2000 SAGE Publications (London, Thousand Oaks and New Delhi),
Vol. 6(3–4): 107–126
[1357–034X(200009/12)6:3–4;107–126;015473]

But why consider a mundane technology such as walking boots? As Borgmann (1984) noted some while ago, these obvious, 'invisible' technologies receive little academic attention yet they are pivotal in shaping everyday life, and do so often through the body (see Falk, 1995; Latour, 1992; Law, 1994). In this article, the aim is to explore how walking boots are, as both material and semiotic resources for the body, instrumental in a variety of relations into which one enters with the natural environment. By way of framing this exploration, I consider two versions of such relations (of course, there are numerous models of such relations – for an important overview, see Macnaghten and Urry, 1998). The first is the Romantic sublime that arguably underpins one particular warrant for environmental action, namely 'environmental expressivism' (Szerszynski, 1996). The second, drawing on ecological psychology, mutually embeds humans and landscape in what Ingold (1993) calls the taskscape. I choose these versions because they sit at opposite ends of a spectrum of the relationality of humans and the natural environment. The sublime assumes a pre-existing separation between human and environment; the two must be brought together, reconnected; mutuality must be re-invented. In contrast, regarding the taskscape, there is a pre-existing, fundamental mutuality between human and environment that shapes both. In both cases, the body is central. The body is either brought to a point where it can be immersed in nature (the sublime), or always already immersed in nature (the taskscape). I argue that in their very different ways (respectively, aspirational and presuppositional) both these perspectives posit a 'pure' relation between humans and the natural environment that neglects the complicating role of technology. As we shall see, both the sublime and the taskscape have been refashioned with the advent of epochal technologies. However, I will focus on the way that local exchanges between humans and environment are fundamentally heterogeneously mediated by mundane technologies, specifically, walking boots.

In what follows, then, I will briefly review, and critique, 'pure' conceptions of the relations between bodies and environments, emphasizing in particular how these relations are mediated by mundane technologies such as walking boots. This point will be used as the basis, in the next section, for an exploration of how the relations between environments and bodies entail both the material and the semiotic – are, in sum, heterogeneous. Drawing upon the philosopher Michel Serres, I will trace some of the ways in which the material and the semiotic interweave – move from one to the other in complex ways. As part of this analysis, I will examine how boots, insofar as they incorporate, for example, all manner of standards, reflect some of the key facets of late modernity, even while in other respects they serve as a ground for the tacit critique of the modern. Finally, in concluding,

I will draw out some of the political issues raised by this way of theorizing the relation between bodies and environments.

The 'Pure' Relation

The Sublime[1]

To experience the sublime in nature is to indulge a sense of astonishment, the mind being so overwhelmed by the natural object that it is, in part, horrified. As Nye (1994) notes, for Burke it was nature itself which triggered this response, whereas for Kant the mind played a much greater part. However, our view of the sublime is mostly shaped by the Romantics. To Wordsworth and Coleridge, the sublime serves in the transcendence of both mind and nature: 'they are attributed a spiritual dimension that is greater than the merely individual and the material' (Day, 1996: 45; also see Bate, 1991; Williams, 1973). Insofar as the sublime still figures in contemporary apprehensions of nature, we can, following Szerszynski (1996), attach it to the rubric of environmental expressivism. According to Szerszynski, such environmental expressivism 'refer[s] to the notion that individuals can reconnect themselves with nature through the recovery of an authentic state of being, one that has been lost due to the artificiality of social existence' (1996: 120). For the expressivist, then, there is an effort to recover 'unmediated experience' of nature: environmentalist expressivism 'is not so much a narrative of us saving nature, but of nature saving us, for only if we abandon modernist notions of control and domination . . . can we know what to do (as environmentalists)' (1996: 121).

Of course, the possibility of the sublime relation to nature rests on a complex and heterogeneous configuration of conditions that entailed, among many other facets: the comprehensive revisioning of nature, so that mountains were no longer ugly by virtue of embodying human corruption and sinfulness but exemplars of natural beauty (e.g. Bowler, 1992; Thomas, 1984); traditional gender divisions reflected in the way that nature and the feminine serve merely to facilitate the male experience of the sublime (Day, 1996); the evolving differentiation between country and city (e.g. Williams, 1973); and the development of new transport systems which enabled greater access to the countryside (e.g. Wallace, 1993). Further, as an example of the contemporary tourist gaze, the sublime is intersected, and 'interfered' with, by many other forms of apprehension – the spectacle, for example (see Macnaghten and Urry, 1998; Urry, 1990, 1995). Moreover, as Nye (1994) has persuasively argued, the object of the sublime has been historically variable, shifting at the height of American industrialization from the natural to the technological. Nevertheless, within this complex, heterogeneous configuration which serves in the constitution and transportation of particular sorts of persons to a particular sort of nature, mundane conditions must hold if the sublime relation is to be realized.

At base, then, it is necessary for the body to be in moment-by-moment contact with the environment in such a way that the sublime is enabled. This moment-by-moment contact can take several forms: there is Wordsworth's familiar and companionable peripatetic, Hazlitt's solitary, contemplative walking, Thoreau's sauntering into the western wilderness (Wallace, 1993). But in each of these cases, both nature and body must behave: nature must not be so violent, that is, so life-threatening as to distract from the sublime; the body must not be in so much pain as to disable contemplation of the nature, so as to 'unmake the world' to use Scarry's (1985) apt phrase. So, what makes the body 'behave', comport itself with such ease and comfort as to be virtually absent from the process of sublime apprehension? Instrumental here are mundane technologies such as walking boots. These mediate the sublime relationship (though, as we shall see, they do this in complex ways) by obviating, screening out, the little discomforts that arise in the moment-by-moment contact between bodies and local nature, so that the grander connections between human mind and sublime nature can be uninterruptedly accomplished.

The Taskscape

Raymond Williams (1973) has traced out in dazzling detail how the sublime relation fits into a complicated literary history that is a reflection of the manifold shifts in the relation between city and country. A key, though much nuanced, feature of this literature is the attempt to recover a pure, Edenic nature. It would seem, Williams tells us, that in every period, there is a literature which harks back to the better, purer, more 'natural' days of life in the country, even though each period may subsequently be identified with just those better, purer, more 'natural' days of life in the country. The sublime relation and environmental expressivism outlined above are fundamentally concerned with the re-entry into – the re-connection with – nature. Moreover, the sublime and the expressivist are essentially *special* relations. They are apprehensions of nature that enlighten spiritually, poetically, politically. By comparison, I now want to consider an account where this connection is pre-existing, indeed, foundational.

According to Ingold there is also a quotidian relation to nature wherein knowledge gained from nature 'is essentially practical: it is knowledge about what the object affords' (1992: 46). Here, Ingold is drawing directly upon Gibson's ecological theory of perception, and especially Gibson's concept of affordance, defined as 'what it (the environment) offers the animal, what it provides or furnishes'

(Gibson, 1979: 127). As such, an animal's environment offers a range of possible actions to it. Crucial here is the idea that such possible actions reflect the capacities and limits of the animal's body. An area of flat ground thus 'affords' a variety of actions – lying, sitting, standing, crawling, hopping, jumping – that mirror the animal's corporeal capabilities. The environment, as a set of surfaces, does not determine an animal's doings, it merely 'suggests' the array of possible doings. Thus, as Heft elegantly excavates, affordances are relational insofar as they are 'the environmental counterparts to the animal's behavioural potentialities' (1989: 6).

Now, these affordances are not oriented toward a passive, sedentary perceiver (as is common in cognitivist, Cartesian models of perception, see Heft, 1989; Michael and Still, 1992), but to an organism that actively and intentionally explores its environment. This on-going activity of animals serves in the limiting of affordances: '[d]epending upon the kind of activity in which we are engaged, we will be attuned to picking up a particular kind of information, leading to the perception of a particular affordance' (Ingold, 1992: 46).

Ingold, along with others (e.g. Costall, 1995; Heft, 1989), extends Gibson's notion of affordance by 'socializing' it. Indeed, affordance becomes intrinsically social insofar as for humans, activity is always culturally invested. As we move through the landscape, the affordances we perceive reflect our circumstances as embodied, cultural beings. As such, for Ingold (1993), the landscape is 'dwelling' in which nature and culture, mind and matter (indeed, evolution and history – see Ingold, 1996a) are indissoluble. This is because landscape is effectively tied up with those who dwell there, their movements, their practices, their workings. To dwell entails a present that embodies the past and projects into the future. This dwelling, Ingold brilliantly shows us, is fundamentally concerned with practices – it is a *taskscape*. Importantly, the taskscape is crucial to the social meaning of landscape; it persists only so long as there are people continuing to practise those activities, those practices of dwelling in the particular landscape. Macnaghten and Urry usefully summarize Ingold's analysis:

> . . . that there are spatially and temporally distributed tasks; that these are organised through a variety of social practices; that relationships with what is taken to be nature are embodied, involving a variety of senses; that there are 'physical' components of walls, textures, land, plants and so on, which partly constitute such 'dwellings'; that the past is continually redefined in terms of the present and projections into the future; that such redefinitions of the past involve forms of collective memory work; and that landscapes are never completed but always subject to contestation and renegotiation, using materials, signs and activities from various pasts as they are projected into diverse futures. (1998:168)

There are faint echoes here of what Williams has called 'the perpetual retrospect to an "organic" or "natural" society' (1973: 96). But Ingold's is not a nostalgia for a historical Arcadia; it is, rather, epistemologically Edenic. Human knowledge is grounded in the indissolubly corporeal and cultural relation to nature – it is a practical, embodied knowledge that is also inscribed in the landscape and vice versa. 'We play our part in a process that is, in effect, the world's formation of itself' as Ingold (1996b: 184) puts it. The 'purity' of this vision lies in its unity, that is, the seamlessness of humans and environment. Yet, integral to this integrity, especially in modern societies, are a huge array of entities that mediate the relation of bodies to nature. Over and above, or rather, beneath and below, those that Macnaghten and Urry summarize above, and expand upon subsequently in their globalization and 'scientification' of the taskscape (e.g. the impact of diverse new spatial practices such mass travel, tourism, photography), are mundane technologies.

As such, affordances are not simply 'between' individual bodies and natural surfaces, but are modulated by other everyday entities – the co-presence of other human and non-human animals certainly, the deployment of sophisticated technologies such as four-wheel drives increasingly, but also such mundane technologies as footwear. These reshape the affordances of nature by expanding the range of possible actions available to the body. Further, walking boots are parts of the environment in themselves insofar as they are composed of surfaces (and, of course, this point applies to any tangible technology). As such they afford as well, and their affordances affect subsequent affordances. When we take such mundane technologies into account, what we begin to see are *cascades* of affordances: for example, socks afford the easier wearing of boots which afford the attachment of crampons which afford the climbing of snow-covered slopes which themselves become 'affordable', that is to say, climbable. Of course, these cascades are not determined: there is no necessity in one artefact articulating with another through such affordance cascades. Indeed, the affordances of any technology are always, at least potentially, ambiguous. In other words, the affordances of technologies – their functions – are, like those that inhere in taskscapes, contestable. What an artefact affords, how it functions is, in principle, plural (see Preston, 1998). Yet, there are constraints to how an artefact functions. These range from the on-going activities and plans of the individual actor, as Ingold points out, through the efforts by manufacturers to prescribe, materially and semiotically, 'proper' use, to the subcultural subversion or re-appropriation of such artefacts.[2]

In light of all this, practical knowledge, as derived in and from the taskscape, which, as we have seen, embodies, and is embodied by, its human occupants, becomes problematic. Or rather, it becomes fragmented, 'impurities' intervene from 'beyond' because, on the one hand, the taskscape itself becomes hugely expanded (as Macnaghten and Urry argue), and, on the other, so does the body as the body-regime (Rose, 1996) spans more and more actors, technologies, relations. Further, this unity and seamlessness that is the taskscape seems always liable to fracture through the sometimes wayward operations of mundane technologies. In addition to mediating relations between humans and humans, and humans and nature, mundane technologies can also disturb these relations, in the process opening up the possibility of new meanings and new relations, that is, a reconfigured taskscape. It is to this complicating role of mundane technology that I now turn via a brief consideration of the pivotal work of Michel Serres.

Quasi-Objects and Parasites

The philosopher Michel Serres has been fundamentally concerned to illuminate the contact and communication between disparate entities and endeavours. The sorts of questions he addresses ask, for example: how does a message move from the realm of the natural into that of the social? What carries this message? How does this message get conveyed or disrupted? Central here is the idea that this message is not simply a matter of meaning, it is also physical – it is heterogeneous (see Akrich and Latour, 1992; Latour, 1992; Serres and Latour, 1995). This is underlined when we consider Serres' concept of the quasi-object.

In our routine rendering of the distinction between objects and subjects, we neglect the heterogeneity and circulation that lies behind subjects and objects. These are, in Serres' terms, really quasi-objects and quasi-subjects. For Serres, quasi-objects are pivotal in the production of society, for what makes the social possible, according to Serres is the movement of quasi-objects:

> Our relationships, social bonds, would be airy as clouds were there only contracts between subjects. In fact, the object, specific to the Hominidae, stabilizes our relationships, it slows down the time of our revolutions. For the unstable bands of baboons, social changes are flaring up every minute. . . . The object, for us, makes our history slow. (1995a: 87; see also Latour and Strum, 1986; Strum and Latour, 1988)

But this movement of quasi-objects is not separate from human relations: 'The relations at the heart of the group constitute their object; the object moving in a multiplicity constructs these relations and constitutes the group. These two complementary activities are contemporaneous' (Serres, 1991: 102; see also Latour, 1992). Some quasi-objects are like jokers, or what Serres also calls, 'blank arms' – they are still undetermined, they might pass between humans but do not necessarily specify the relations. As Serres (1982: 232) remarks: 'Our quasi-objects have increasing specificity.' In other words, as the specificity and particularity of quasi-objects become greater, so this mediates, and is mediated by, the increasing specificity and particularity of social relations (see also Serres, 1995b).

Now, in the relation between humans and the natural, quasi-objects once again mediate. The 'messages' – the flows of signs and energies – between humans and natures pass through mundane technologies. The boots that enable people to reach the sublime wherein something like an intersubjective relation is realized, the boots that wear a path through the landscape and recapitulate its meaning as a taskscape – these boots 'circulate' in the immediacy of contact between bodies the environment, inscribing the former with the latter and vice versa.

Yet, these boots are not simple intermediaries, going about their business as innocent conduits, pristine channels. They too contribute to this process of communication – this exchange of meanings – by introducing their own heterogeneous messages. Or rather, they can act as interference. We can think this through Serres' (1982) notion of the parasite. The parasite has several meanings, for example, an organism that takes from its host without giving anything in return, or a disrupter of a signal between communicator and receiver. In all these, the possibility of pure communication (in the heterogeneous sense of a passing of signs and materials – Akrich and Latour, 1992) rests on an exclusion – a bracketing, a removal of those entities, processes, parasites that would otherwise disturb the connection. Yet these parasites (one of which is nature itself – see Serres, 1995c) persist. In 'The Parasite' Serres (1982) explores in subtle detail the variations of form that the 'parasite' can take through an extended meditation on the uninvited guest at the dinner table who exchanges stories for food. Here, we see how stories (the communicational, the semiotic) are transformed into the material (food, shelter) and vice versa. Serres is, in fact, tracing the transitions from the material to the semiotic and back again, charting the ways in which material interjections enable the emergence of new meanings; how the play of meanings generates new material arrangements. The movement of materials (food) at the dinner table is redirected through the uninvited guest's stories; the flow of signs (conversation) at the dinner table is reconfigured, through the uninvited guest's noisy eating.

Boots are invited, indeed, sometimes necessary guests in the heterogeneous dialogue between humans and the environment. They at once mediate this dialogue, and transform it. They are simultaneously intermediaries, faithfully transferring messages to and fro, and re-shaping those messages. This ambiguous, or rather dual, aspect of walking boots will be discussed in the sections that follow. However, before we go on to this, it is important to stress that the account I develop of the multifarious role of walking boots – a role that is, to reiterate, heterogeneous and distributed, as we shall see – is not primarily, let alone exclusively, phenomenological. I am not concerned with the experience of nature per se by the walker. Rather, I am more interested in tracing out, following Serres, the movement of messages – material and semiotic – between human and natures. To be sure, the phenomenological comes into this, but I am also keen to emphasize that the 'natural' is also in receipt of these messages. The point, then, is to explore, the possible meanings that the relation between humans and environment can hold once we take into analytic account the mediations of the walking boots as an important adjunct to the human body.

Talking Boots and Walking Bodies

So, we are assuming the mediating role of mundane technology such as walking boots (other forms of clothing would do just as well, of course, though entailing different circuits of messages). They are intrinsic to our supposedly pure relations to the environment, incorporating and expanding affordances. But, as argued in the preceding section, they are not simple means of conveyance, or conduits by which messages (defined as heterogeneous) pass from humans to nature and back again. They intervene and influence these messages and they do so in heterogeneous ways. I will consider four such interventions: first, there is the role of boots as mechanical technologies that can cause pain, dissolving identity and the relation between humans and nature; second, there is the role of boots as signifying style and identity; third, there is the role of boots as embodiments of procedures of standardization and objectification; and finally, there is the role of boots as technological means of physical and ecological damage to nature. I shall also demonstrate that these aspects of walking boots are never distinct – that they overlap and 'recapitulate' one another in various ways.

Pain

Walking boots are, ideally, meant to be invisible, or, rather, intangible. They conduct one to and from and through the environment without being noticed. They should be 'broken in' – fit one's feet comfortably so that contact with the ground is phenomenologically unproblematic. In such instances, boots are simple tools that quietly expand the capacities of the body, and thus the affordances of nature.

Yet, sometimes, they do not work in this way. Sometimes they are painful. They might not broken in yet – the material being relatively unyielding, or still to be contoured to the shape of the foot. Or they might simply be too small through a miscalculation at the time of purchase, or because of changes in one's feet under different conditions (for example, slight swelling in a hot climate might make the boots, which fitted perfectly in the relative coolness of an outdoor clothing store, half a size too small). At such points, the boots become a version of Serres' parasites – they materially intervene in what should be a smooth flow of communication between nature and body; they disrupt, abbreviate, curtail the signals, or materials, that pass between these two entities. Instead of being silent, they constitute noise which drowns out, or rather, which utterly interrupts, many of the usual flows between humans and nature (though obviously not such material flows as the exchange of oxygen and carbon dioxide). One can no longer experience the sublime or realize all the potential affordances enabled by the boots, for they are no longer, strictly speaking 'walking boots'. This is because, to be in such pain, to experience such agony leads to a distanciation of the 'self' from the agonized body (the self is somehow 'outside' the body) or the fragmentation of self such that there is, as Bendelow and Williams (1998: 139) put it: 'a "psychic splintering" and "disintegration", devoid of content, entirely cut off from the surrounding sociocultural world'. In this case, of course, that sociocultural world is the local environment through which one is travelling. To reiterate Elaine Scarry's (1985) phrase, there is an unmaking of the world. The physical interventions of the boots have the effect of breaking up the semiotic flows between nature and human – it becomes very difficult, if not impossible, to grasp the sublime under such corporeal circumstances. They also alter material flows – the contacts between body and environment become modified, that is, constrained as movements become more circumspect in the process of trying to modulate each impact of foot on ground, and thus minimize the pain. One might say that one is forced to tread lightly upon the Earth.

However, this is not all that happens. The process of being in pain can also lead to seeking out ways of distracting oneself, or of reconnecting with the surrounding sociocultural world. Other walkers' chit-chat, behaviour, comportment can all serve as a means of drawing oneself out of one's agonized body, and back into contact with the world (see Bendelow, 1998; Bendelow and Williams, 1998). These are semiotic parasites upon the parasite of the boot – they intervene to break up its breaking up of the communication between human and local environment. But, we must ask, what constitutes a set of signs so powerful as to break into and, perhaps only momentarily, disrupt this flow of pain? Inevitably, these will be various. There are such signs as, say, the immediate dangers posed by nature (an earth tremor, say). Yet, there are also human significations which can intervene – it is not simply any old chit-chat or behaviour or comportment that can be 'powerful' enough to be a, for want of a better term, 'counter-parasite'. There must be something salient about it: perhaps, it relays another's even greater agony; or perhaps it contravenes some deeply held value about proper comportment in the environment. The details, in the present context, are not important. The point is that human others can interrupt the interruption of pain by virtual of their proximity in space and time. The material pathology of this body–technology–nature nexus is, as it were, treated by signs.

There is another aspect to this 'counter-parasitism': instead of proximity, there is prospect and distance. Williams and Bendelow (1998) note that pain can be narrativized; pain can thus become meaningful (say, in terms of the discourses of gender difference in 'coping' with pain), productive (say, in that pain narratives serve in the revaluation of lifestyle). In the case of agonized walking, we can say that the prospect of narrativization – of generating signs about the pain for sharing with relevant others – can also be counter-parasitical. The prospect of telling the story of pain and suffering can thus be a distraction. It is the prospect of telling projected into the future and into a distal space. Thus, one again removes oneself from the immediacy of the pain. But to do this one must draw on a stock of narratives – the heroic, for example. One was in pain, yet overcame it (and no doubt there are many variations on this depending on the subcultures to which one is attached: it is not unimaginable that some climbers use these narratives as a mark of prowess).

This discussion has not been concerned merely with the seemingly marginal case of pain. Rather, the issue of pain in the context of human–nature relations signifies the importance of embodiment, particularly in relation to the inter-objectivity of these relations. That is to say, if boots are normally intangible, when they become painful, when they become noise, they throw into relief the inter-objectivity of human bodies and natural environment. But, even when we see this inter-objectivity so clearly, we also see that it is in fact cross-cut by the cultural – the interventions of what we might call 'proximal' and 'distal' narrativizations.

Style and Identity

The preceding section has, in the first instance, treated walking boots as material objects that intervene in, and facilitate, the flow of communication between humans and environment. However, boots are also the objects of semiotic consumption – they are means to distinction in Bourdieu's (1984) sense signifying the sort of walker one is (see also, for example, Featherstone, 1991; Lury, 1996; Slater, 1997). In other words, certain boots, such as Caterpillar or CAT boots, signify urban walking – they are associated with certain groupings within urban youth. In contrast, 'serious' walking is signified by such brands as Berghaus, Meindl, Salomon, Zamberlain or Scarpa.

When one considers the ways in which walkers themselves account for their boots, then this aesthetic or stylistic aspect of walking is routinely denied. For example, when members of the Lancaster University Hiking Club were asked about the ways in which they chose their walking boots, they said, and underlined, that it was, in essence, a functional or practical matter. The boots were chosen on the basis that they did their job – that is, allowed them to walk in comfort (though additionally, they also mentioned that boots were chosen on the basis of their durability and their price; aesthetic attractiveness was a secondary matter, a pleasant bonus). A similar set of criteria was mobilized by contributors to an Internet survey of walking boots choices ('Walklist': http://www.Teleport.com/~walking/shoepoll.htm – 18 May 1998). For example, in response to the question 'What do you look for in a walking shoe?', one finds the following sorts of answers: 'I look for comfort, durability and price'; 'Although I am easy on shoes, I still want comfort and durability'; 'I look for shoes that are very light, good flexibility and plenty of room for the front part of the foot, i.e., the toes'; 'I look for footwear that is well cushioned, contoured to the shape of my feet, and which provides a maximum of ventilation'.

These practical criteria are based on a tacit discourse of 'needs'. The relative uninterest in style however is not a mere absence. It too signifies – the 'serious walker' should not be concerned with the 'trivialities' of aesthetics or style. This denial of the aesthetic dimension marks difference from those who are concerned with style, and who are thus not serious about walking. One might say, drawing on Sharon Traweek's (1988) notion in relation to physicists who practise a 'culture of no culture' (and Haraway's reworking of this as the 'nature of no nature'), that we are witnessing the 'fashion of no fashion'.

Further, this 'fashion of no fashion' has implications for the way that pain is to be narrated. For the serious walker it has to be a matter of a mistake or a miscalculation in the purchase and 'breaking in' of boots (after all, there is plenty of instruction available on how to look after and 'customize' one's boots in the walking literature and on the web). Pain cannot be the upshot of sacrificing functionality for fashion. Once again, when we consider something as seemingly basic as pain (as a flow of signals between environment and body), we can see that it can be parasitized by the significations of the boot. Ironically, the boot as an artefact of consumption signifies identity and social differentiation through the very repudiation that it is such an artefact.

Standardization and Objectification

If the walking boot signifies, at least for some, the 'fashion of no fashion' and a lack of concern for aesthetics even while it enables access to the environment as, among other things, an aesthetic setting, it also signifies the very obverse. Walking boots, though on one level mundane technologies, are, on another, highly sophisticated: they incorporate complex materials, are constructed through sophisticated techniques, are standardized through intricate processes of measurement. In other words, even as they take us out of modernity to experience the sublime, or attain seemingly basic affordances, they carry with them the very essence of modernity. At the very heart of human–nature relations as they are played out in the simple act of walking, are the modern practices of objectification and standardization. To put this point in more general terms, the pure relation is not simply localized between human individual body and immediate environment – it incorporates the global: boots mediate between distributed heterogeneous networks that encompass globalized systems and the global environment.

Boots, as the medium between bodies and natures, are, I have argued, not innocent. They can intervene as parasites in those cases where they 'cause' pain. But, if part of what they are mediating is a flow of signs that sustains the expressive, and that enables the sublime, they also interject other signs, those of the modern. Boots are made of materials – leather, plastics, rubbers – that entail in their manufacture and combination the objectification of nature. Further, in the production process these materials must be standardized: measures such as grams and volts and centimetres must be kept constant. This requires the calibration of production technologies against other machines that 'hold' a given unit or measure, which must themselves be calibrated and so on *ad infinitum* in what O'Connell (1993; see also Bowker and Star, 1996) calls the circulation of particulars (that is, those machines that do the calibrating). At stake here is the fact that the pure relation (epistemological or expressive) mentioned above is only possible because of the 'purification' associated with modernity (Latour, 1993a), which ensures elementary characteristics: boots are the right size and stay that size; boots are not toxic; boots do not lose their flexibility or fall apart when exposed to certain conditions, and so on and so forth. Behind this purity lie heterogeneous networks which ensure it.

To reiterate, this objectification (and I have only mentioned that of materials, not of the humans who are involved in the manufacture of these garments) is globalized. In serving to signify objectification, walking boots also signify the distributed, global processes of their own production and dissemination. The 'intimacy' of the sublime and of affordance is structured by the global. As writers such as Beck (1992) and Giddens (1991) have argued, inherent here are the processes of trust in (or, rather, ambivalence toward) the abstract systems that monitor standards and ensure their implementation. As such, the irony is that that the embeddedness in nature that affordance and the expressive signal is only made possible by our disembeddedness. However, this process of disembedding can link up to issues of consumption and identity. Sometimes, for example, there is a de facto re-embedding in a subculture that attempts to move beyond standardization, to bespoke, as it were, standardized boots. Thus, for example, *The Great Outdoors* magazine organized a 'boot camp' where experts would be on site to advise walkers on fit, feet and boots. These experts are not the same as expert systems – they are represented as artisans, attached to the subculture of walkers and hikers. So, sometimes the irony is itself ironized; we see cascades of embeddedness and disembeddedness and re-embeddedness.

If this irony of the necessity of disembeddedness in the realization of embeddedness renders walking boots such a good conduit between humans and the environment, it is also the case, once more, that boots parasitize on that relation in another way – as the product and means of environmental damage. It is to their part in the despoliation (and repair) of the environment that I now turn.

Damage and Repair

We do not have to rehearse the litany of destruction of, and damage to, the environment that underpins the production of walking boots. Suffice it to say, that they are caught up in the usual unsustainable uses of resources and energy that characterize the modern world: where once one could have enjoyed the sublime, say in the rainforests of Indonesia, now there are rubber plantations. As such, they enter another message into the communication loop between humans and environment – they are, as manufactured products of modernity, at once media for the re- and mis-alignment of humans and nature.

But there is another, more immediate, way in which walking boots do harm. If boots are instrumental in the damage wrought on feet, they also play a part in the damage wrought on nature. They can also directly change the environment by, for example, eroding landscape features (let alone pathways), abrading plant-life or importing foreign species. As the body and nature are brought together, the latter becomes endangered (although so does the former increasingly – but that's another discussion). Now clearly this is a matter of the density and frequency of walking boots at any particular site. Boots once again parasitize the pure relation by interjecting a different set of messages that stem from their objectification in, for instance, visitor statistics to a given site.

Yet, individuals can also mitigate their personal impact upon the environment. They might choose boots designed to minimize erosion, boots such as The Brasher Boots Company's Hillmaster and Lady Classic whose three-layer sole has an outsole 'Studded and cleated to provide maximum traction whilst design minimizes clogging thus improving safety and reducing footpath erosion' (http://www.spelean.com.au/BRA/Hillmaster.html). Or else, in relation to the more global implications of resourcing, production, distribution and waste, they might check on the manufacturers to ensure that they have an environmental ethic and audit in place, perhaps practising life-cycle assessment techniques to ensure that, from 'cradle to grave', their products are as environmentally sustainable as possible (Molloy, 1997a, 1997b). Here again, there is recourse to standardization and objectification.

But, again ironically, walking boots are also a tool for repairing damage. For instance, they serve in the conveyance of bodies that make the local measurements of habitat loss and that document the decline in populations of flora and fauna. Yet again, the processes of standardization and objectification come into play. Further, the boots convey the bodies that do the repairing and safeguarding – that replant, that put up signs warning and directing walkers, that rebuild walls – that, in sum, conserve the landscape.

The pure relation is, in still another way, infected by other messages that bear, among others, the marks of consumption and standardization.

Concluding Remarks: The Politics of Multivalence

In exploring the 'pure' relation (in its two guises) between humans and the natural environment, I suggested that we need to take into account the role of mundane technologies. Walking boots, which would seem to mediate the relation between humans and environment, are, we see, not mere simple conduits or intermediaries. The cycles of communication between human bodies and nature are, in a multitude of ways, parasitized by the interventions – semiotic and material – that a mundane technology such as walking boots can manifest. As quasi-objects they are not faithful messengers, but interject their own messages. Even so a peremptory a meditation (if that's not an oxymoron) on the role of walking boots in the relations between human bodies and the environment, as presented here, has quickly yielded a morass of interleaving messages that is largely hidden in the pure relation. The foregoing discussions of pain, style and identity, standardization and objectification, and environmental damage and repair have spanned the inter-subjective, the inter-objective and various permutations thereof. The semiotic is cross-cut by, and converted into, the material and vice versa (and the same goes for the real and the socially constructed – see Latour, 1997). Where walking boots mediate an inter-subjective relation to nature – something akin to the expressive, say – they also parasitize on this, curtail it by virtue of entailing objectification on a multitude of levels (for example, as mediators of erosion). In the process of putting oneself in the position of experiencing the sublime, one also 'does' consumption, damage, standardization, disembeddedness and so on, and innumerable more or less subtly nuanced versions of these. To speak of the sublime, is simultaneously to speak of standardization, consumption, damage and repair. And to refer to any of these terms is tacitly to refer to the others. Nature does not only afford for bodies – the underlying assumption in Ingold's profound analysis – but affords to bodies-and-boots (and conversely, one doesn't walk in 'nature', but in 'nature-and-boots'). Once walking boots are taken into account, we see not simply a process of practice and perception, but interaction and communication into which are imported a vast array of other messages. Here, I should emphasize that the focus on walking boots is not meant to imply that all relations between humans and nature entail these artefacts. Clearly, this would be ridiculous. Rather, walking boots have served as a figure for exploring the role of mundane technologies more generally, in particular the necessary interventions of such technologies in this relation, and the complexity of this intervention. Obviously enough, other such technologies will yield different configurations of complexity. Nevertheless, the general point holds, namely, that on accepting the co-presence of these technologies with bodies and natures, then the relations between the latter must be seen as multiplicitous – highly complex and contradictory.

Now, the multiplicity of these messages suggests that walking boots are like blank arms or jokers, in Serres' terms. They do not define the relationship – rather, they show how the pure relationship, or, for that matter, any relationship between human and nature is multivalent. Where does this leave us politically? More grandiosely, we can ask: what are the politics of multivalence? On one level, the analysis presented above is somewhat opaque when it comes to guiding a politics in relation to environmental damage, loss, risk, etc. It operates at so mundane a level, at which there are so many variants of the human–boot–nature combination, that there can be no overarching principles by which to derive 'what to do' as an environmentally concerned actor. However, this is the point. It is an analytic that 'advises' modesty and caution in the derivation of such principles (see Latour, 1993a, 1998).

Yet, on another level, what this analysis has implicitly begun to sketch are a series of rough portraits of particular hybrids. That is to say, particular combinations of human/mundane technology/nature have been hinted at: the hybrid walker with environmentally friendly boots who is mindful of their position as a 'tourist' versus the walking hybrid who romanticizes nature while being oblivious to the complexities of its comportment within nature. To stress, we see these hybrids as entities in their own right with properties different from their individual parts (see Latour, 1993b). Latour (1993a) asks us to consider modernity as fundamentally concerned with the purification of hybrids – their separation into humans and non-humans. All the while, these hybrids are proliferating unjudged. What we need, then, are ways of judging the value of hybrids in their specificity: some are good, some are bad (see Michael, 1998, for an example of this). But as we have seen, the massive, heterogeneous distributedness of these combinations makes it difficult to identify the borders of a hybrid (Callon and Law, 1995). To ask ' "what" is being judged, assessed, evaluated?' is also to ask ' "who" is drawing the boundaries and "what" is rendering judgement assessment, evaluation'? For are 'we' not ourselves hybrid? Perhaps, we can posit forms, loci or figures of judgement that are themselves distributed over configurations, arrangements of entities and relations (see Michael, in press). Once more, such complexities and contingencies suggest, as Latour sketches, a sort of cautious ecopolitical strategy which appeals neither to a pure nature nor a pure politics, but rather puts 'in place other procedures for politico-scientific research and experimentation' (1998: 236) in which the indissolubility of the social and the natural is key.[3] These procedures (and their associated vocabularies, practices and relations) have barely begun to be demarcated. Whatever form, if any, they eventually take, we can be sure that walking boots will have a part to play.

Notes

1. As should be clear, I use the term 'sublime' in the restricted sense of the Romantics, though this movement was hardly coherent and the meaning attached to this term hardly uniform (see Garrard, 1998). Further, I do not pursue here any of the poststructralist re-workings of the sublime, for example, those of Lyotard (e.g. 1991) which, in any case, have not been used especially to theorize environmental concern, as Conley's (1997) survey of the relation of contemporary French theory to ecopolitics suggests.
2. For examples of such contestations of function in the sociology and anthropology of technology, see Bijker and Law (1992), Lemonnier (1993) and Lie and Sorensen (1996).
3. As various authors have noted, there are dangers to such an 'uncertain' hybridic strategy not least in its seeming relativism. As examples of critiques of the political implications of such heterogeneous figures, specifically of Haraway's comparable concept of the 'cyborg', see Prins (1995), Lykke (1996) and Hess (1995).

References

Akrich, M. and B. Latour (1992) 'A Summary of a Convenient Vocabulary for the Semiotics of Human and Nonhuman Assemblies', pp. 259–63 in W.E. Bijker and J. Law (eds) *Shaping Technology/Building Society*. Cambridge, MA: MIT Press.

Bate, J. (1991) *Romantic Ecology: Wordsworth and the Environmental Tradition*. London: Routledge.

Beck, U. (1992) *The Risk Society*. London: Sage.

Bendelow, G. (1998) 'Painful Bodies: The Role of the Natural', paper presented at the British Sociological Association 'Making Sense of the Body' Conference, University of Edinburgh.

Bendelow, G. and S.J. Williams (1998) 'Emotions, Pain and Gender', pp. 253–67 in G. Bendelow and S.J. Williams (eds) *Emotions in Social Life: Critical Themes and Contemporary Issues*. London: Routledge.

Bijker, W.E. and J. Law (eds) (1992) *Shaping Technology/Building Society*. Cambridge, MA: MIT Press.

Birke, L. (1999) *Biological Bodies and Feminism*. Edinburgh: Edinburgh University Press.

Borgmann, A. (1984) *Technology and the Character of Contemporary Life*. Chicago, IL: University of Chicago Press.

Bourdieu, P. (1984) *Distinction*. London: Routledge and Kegan Paul.
Bowker, G.C. and S.L. Star (1996) 'How Things (Actor-net)work: Classification, Magic and the Ubiquity of Standards', Internet Document [http://alexia.lis.uiuc.edu/~bowker/actnet.html].
Bowler, P.J. (1992) *The Fontana History of the Environmental Sciences*. London: Fontana.
Callon, M. and J. Law (1995) 'Agency and the Hybrid *Collectif*', *South Atlantic Quarterly* 94: 481–507.
Conley, V.A. (1997) *Ecopolitics: The Environment in Poststructuralist Thought*. London: Routledge.
Costall, A. (1995) 'Socializing Affordances', *Theory and Psychology* 5: 467–81.
Day, A. (1996) *Romanticism*. London: Routledge.
Falk, P. (1995) 'Written in the Flesh', *Body & Society* 1: 95–105.
Featherstone, M. (1991) *Consumer Culture and Postmodernism*. London: Sage.
Garrard, G. (1998) 'The Romantics' View of Nature', pp. 113–30 in D.E. Cooper and J.A. Palmer (eds) *Spirit of the Environment: Religion, Value and Environmental Concern*. London: Routledge.
Gibson, E.E. (1979) *The Ecological Approach to Visual Perception*. Boston, MA: Houghton Mifflin.
Giddens, A. (1991) *Modernity and Self-Identity*. Cambridge: Polity.
Heft, H. (1989) 'Affordances and the Body: An Intentional Analysis of Gibson's Ecological Approach to Visual Perception', *Journal for the Theory of Social Behaviour* 19: 1–30.
Hess, D. (1995) 'On Low-Tech Cyborgs', pp. 371–7 in C.H. Gray (ed.) *The Cyborg Handbook*. New York: Routledge.
Ingold, T. (1992) 'Culture and the Perception of the Environment', pp. 39–56 in E. Croll and D. Parkin (eds) *Bush Base: Forest Farm – Culture, Environment and Development*. London: Routledge.
Ingold, T. (1993) 'The Temporality of the Landscape', *World Archeology* 25: 152–74.
Ingold, T. (1996a) 'Situating Action, V: The History and Evolution of Bodily Skills', *Ecological Psychology* 8: 171–82.
Ingold, T. (1996b) 'Situating Action, VI: A Comment on the Distinction Between the Material and the Social', *Ecological Psychology* 8: 183–7.
Latour, B. (1992) 'Where are the Missing Masses? A Sociology of a Few Mundane Artifacts', pp. 225–58 in W.E. Bijker and J. Law (eds) *Shaping Technology/Building Society*. Cambridge, MA: MIT Press.
Latour, B. (1993a) *We Have Never Been Modern*. Hemel Hempstead: Harvester Wheatsheaf.
Latour, B. (1993b) *On Technical Mediation: The Messenger Lectures on the Evolution of Civilization*. Ithaca, NY: Cornell University, Institute of Economic Research: Working Papers Series.
Latour, B. (1997) 'A Few Steps Toward an Anthropology of the Iconoclastic Gesture', *Science in Context* 10: 63–83.
Latour, B. (1998) 'To Modernise or Ecologise? That is the Question', pp. 221–42 in B. Braun and N. Castree (eds) *Remaking Reality: Nature at the Millennium*. London: Routledge.
Latour, B. and S. Strum (1986) 'Human Social Origins: Oh Please, Tell Us Another Story', *Journal of Social and Biological Structures* 9: 169–87.
Law, J. (1994) *Organizing Modernity*. Oxford: Blackwell.
Lemonnier, P. (ed.) (1993) *Technological Choices: Transformation in Material Culture since the Neolithic*. London: Routledge.
Lie, M. and K.H. Sorensen (eds) (1996) *Making Technology Our Own? Domesticating Technologies into Everyday Life*. Oslo: Scandinavian University Press.
Lury, C. (1996) *Consumer Culture*. Cambridge: Polity.
Lykke, N. (1996) 'Between Monsters, Goddesses and Cyborgs: Feminist Confrontations with Science', pp. 13–29 in N. Lykke and R. Braidotti (eds) *Between Monsters, Goddesses and Cyborgs: Feminist Confrontations with Science, Medicine and Cyberspace*. London: Zed Books.
Lyotard, J.-F. (1991) *The Inhuman*. Oxford: Polity Press.
Macnaghten, P. and J. Urry (1998) *Contested Natures*. London: Sage.
Martin, E. (1989) *The Woman in the Body*. Milton Keynes: Open University Press.
Michael, M. (1998) 'Co(a)gency and Cars: The Case of Road Rage', pp. 125–41 in B. Brenna, J. Law and I. Moser (eds) *Machines, Agency and Desire*. Oslo: TVM.

Michael, M. (in press) 'Dreams of Disembodiment: The Secret History of the Remote Control', in Sarah Cunningham-Burley (ed.) *Exploring the Body*. London: Macmillan.
Michael, M. and A. Still (1992) 'A Resource for Resistance: Affordance and Power-Knowledge', *Theory and Society* 21: 869–88.
Molloy, E. (1997a) 'Product Profile Analysis (PPA)', paper presented to the ERP Environment 'Eco-Management and Auditing Conference', UMIST, Manchester.
Molloy, E. (1997b) 'Social, Ethical and Environmental Accounting: Integrating Values into Management', paper presented to the ERP Environment 'Business Strategy and the Environment Conference', University of Leeds.
Morse, M. (1994) 'What do Cyborgs Eat? Oral Logic in an Information Society', pp. 157–87 in G. Bender and T. Druckrey (eds) *Culture on the Brink: Ideologies of Technology*. Seattle, WA: Bay Press.
Nye, D.E. (1994) *American Technological Sublime*. Cambridge, MA: MIT Press.
O'Connell, J. (1993) 'Metrology: The Creation of Universality by the Circulation of Particulars', *Social Studies of Science* 23: 129–73.
Preston, B. (1998) 'Why Is a Wing Like a Spoon? A Pluralist Theory of Function', *Journal of Philosophy* 95: 215–54.
Prins, B. (1995) 'The Ethics of Hybrid Subjects: Feminist Constructivism According to Donna Haraway', *Science, Technology and Human Values* 20: 352–67.
Rose, N. (1996) *Inventing Our Selves: Psychology, Power and Personhood*. Cambridge: Cambridge University Press.
Scarry, E. (1985) *The Body in Pain: The Making and the Unmaking of the World*. New York: Oxford University Press.
Serres, M. (1982) *The Parasite*. Baltimore, MD: Johns Hopkins University Press.
Serres, M. (1991) *Rome: The Book of Foundations*. Stanford, CA: Stanford University Press.
Serres, M. (1995a) *Genesis*. Ann Arbor: Michigan University Press.
Serres, M. (1995b) *Angels: A Modern Myth*. Paris: Flammarion.
Serres, M. (1995c) *The Natural Contract*. Ann Arbor: Michigan University Press.
Serres, M. and B. Latour (1995) *Conversations on Science, Culture and Time*. Ann Arbor: Michigan University Press.
Shilling, C. (1993) *The Body and Social Theory*. London: Sage.
Slater, D. (1997) *Consumer Culture and Modernity*. Cambridge: Polity.
Strum, S. and B. Latour (1988) 'Redefining the Social Link: From Baboons to Humans', *Social Science Information* 26: 783–802.
Synnott, A. (1993) *The Body Social*. London: Routledge.
Szerszynski, B. (1996) 'On Knowing What to Do: Environmentalism and the Modern Problematic', pp. 104–37 in S. Lash, B. Szerszynski and B. Wynne (eds) *Risk, Environment and Modernity*. London: Sage.
Thomas, K. (1984) *Man and the Natural World*. Harmondsworth: Penguin.
Traweek, S. (1988) *Life Times and Beamtimes: The World of High Energy Physicists*. Cambridge, MA: Harvard University Press.
Turner, B. (1992) *Regulating Bodies*. London: Routledge.
Turner, B. (1996) *The Body and Society*, 2nd edn. London: Sage.
Urry, J. (1990) *The Tourist Gaze*. London: Sage.
Urry, J. (1995) *Consuming Places*. London: Routledge.
Virilio, P. (1995) *The Art of the Motor*. Minneapolis: University of Minnesota Press.
Wallace, A.D. (1993) *Walking, Literature and English Culture: The Origins and Uses of Peripatetic in the Nineteenth Century*. Oxford: Clarendon Press.
Williams, R. (1973) *The Country and the City*. London: The Hogarth Press.
Williams, S.J. and G. Bendelow (1998) 'In Search of the "Missing Body": Pain, Suffering and the

(Post)modern Condition', pp. 125–46 in G. Scambler and P. Higgs (eds) *Modernity, Medicine and Health*. London: Routledge.

Mike Michael is senior lecturer in the Sociology Department at Goldsmiths College, University of London. He has published extensively in a number of fields, including the public understanding of science, the animal experimentation controversy and critical social psychological theory. More recently his interests have turned to the role of mundane technologies in social ordering and disordering. He is the author of *Constructing Identities* (Sage, 1996) and *Reconnecting Culture, Technology and Nature: From Society to Heterogeneity* (Routledge, 2000).

Naked as Nature Intended

DAVID BELL AND RUTH HOLLIDAY

Naked as Nature Intended

> A healthy, brown body beside the blue lake, in the green of the forest, on the mountain-tops: nothing on earth is more splendid. How it shines, how it shimmers against its background, how truly it is the crowning glory of creation! All who have seen it, seen it with their whole souls, know why human beings are not born with clothes on – they know that man is a creature of light. (Zimmermann, 1927, quoted in Szeemann, 1997: 3)

The practices and discourses of the contemporary naturist movements in the West – and those to be uncovered in their histories – are at least in part about articulating a particular embodied relation to nature. While 'social nudity' in this context need not be linked to or enacted in the natural landscape, the evolution and current manifestations of naturism nevertheless centre on a 'philosophy' which is all about bodies in nature. As we shall see, however, the practices of these unclothed bodies in nature have come to be defined and understood within particular frames, through a complex negotiation of discourses of nature, human nature, the natural body and the natural landscape. This negotiation is perhaps most vividly manifest in the ways in which naturism as a practice is positioned in relation to the erotic, and in the counterposing of nature (positively invested) with culture or civilization (negatively invested).

Our article is an attempt to explore these themes. We will begin by sketching the historical development of some Western naturist movements, in an attempt to map their ideological imaginings of nature and the 'natural body'; we will also discuss the ways in which the UK naturist scene at the end of the 20th century is addressing both its inheritance and its current positioning within British society. The issue of naturism's problematic relation to sex, which is central to that

Body & Society © 2000 SAGE Publications (London, Thousand Oaks and New Delhi),
Vol. 6(3–4): 127–140
[1357–034X(200009/12)6:3–4;127–140;015474]

positioning, leads us into a consideration of the erotics of nature, and to a brief look at related practices which also enact an embodied interconnection with nature – those found in outdoor sex, where nature forms the backdrop for 'natural passions'; and in the so-called mythopoetic men's movement, where nature forms the backdrop for 'natural masculinity'. As an expressive and located bodily practice, then, naturism allows us to think about particular bodies, situated in particular constructions of nature.

(Natural) Histories of Naturism

The histories of naturist movements in the West bear common traits, but these are often inflected by particular national (and sometimes nationalist) ideologies and cultural attitudes. A comparative reading of German and English movements as they emerged and interrelated in the early part of the 20th century usefully illustrates this, and will aid us in understanding the concerns of the contemporary English movement. Running alongside these self-described 'movements', which have become institutionalized and bureaucratized across the course of the century, there are other practices of social nudity which we must also consider, such as the nude beach scene which emerged in the USA in the 1960s, partly under the influence of the hippie movement. In the UK naturist scene in the 1990s, there is an ideological divergence or split between naturism and nudism – with the latter dismissed by the former as leftists, radicals or mavericks – which in some ways bears the traces of this divergence in the meaning attached to removing one's clothes in a public, social setting.

The emergence of Western naturist movements in the early 20th century has to be located within a broader cluster of cultural re-orientations towards nature. The aesthetic and romantic appreciation of the countryside, which in Britain emerged from Romanticism, spilled over into a number of embodied engagements with the natural landscape at a time when mass urbanization and industrialization were affecting the nation's population – and its psyche – to an unprecedented degree. Landscape painting and romantic poetry became translated into popular leisure pursuits such as hiking and camping as transport systems opened the countryside up to the urban masses (Bunce, 1994). The 'idyll-ization' of the rural landscape lent it a symbolic role that accounts for this appeal, and tied it into a national discourse that effectively constructed a moral topography of the country. As David Matless shows in *Landscape and Englishness* (1998), the ideological rebonding of the nation, its citizens and the natural landscape was worked through across a number of cultural domains. At the same time as the countryside became an object of mass enjoyment, of course, came calls for its protection and preservation – and, as Matless points out, the emerging naturist movement became caught in the midst of these competing claims about the countryside's role in what he names, after rambling activist and preservationist Cyril Joad, 'the art of right living'.

The art (or arts) of right living Joad enthused over were, first and foremost, technologies of national citizenship, played out across both body and landscape. They evolved at least in part from late 19th-century interests in nature and the rural, which came to be articulated through a number of movements sharing a desire to rediscover (or invent) ways of living perceived as having been eroded by modernity. This 'back to the land' milieu in Victorian England, so eloquently described and assessed in Jan Marsh's book of that name (1982), included movements and individuals with passions as diverse as vegetarianism, folk songs, peasant crafts, land reform, theosophy, communal living and rational dress. Occasionally, the latter also involved an interest in *rational undress*, as the state of the body and its place in nature came under scrutiny. By the inter-war period, as Matless (1995: 93) writes, '[a] particularly landscaped version of English citizenship emerged'; this was simultaneously a particularly *embodied* version of English citizenship, worked up through the disciplines of mass physical culture and exposure to elemental nature. Here is Joad himself, looking to the countryside for its educational, uplifting, humanizing forces:

> Whence can we derive . . . an education alike of body, of mind and of spirit, so happily as from Nature? The feeling of the air upon the skin, of the sun upon the face; the tautening of the muscles as we climb; rough weather to give us strength, blue skies and golden sunny hours to humanize us – these things have their influence upon every side of our being. (Joad, 1934: 150, quoted in Matless, 1995: 108)

Although this quotation derives from the preservationist literature (it is from *A Charter for Ramblers*), it chimes so clearly with the rhetorics of advocates of naturism that the slide between philosophies seems slight. In Germany, indeed, the arts of right living much more readily embraced mass naturism, especially in the youth movements; it is estimated that, prior to its outlawing by Hitler, there were as many as 3 million social nudists in Germany in the 1930s (Anthony, 1998). As we shall see later, the fact that naturism did not seem to fit quite so squarely with the English ruralists tells us some revealing things about national morality, attitudes to nudity and its relation to sex – questions that continue to haunt the naturist movement in the UK.

As with all social movements, naturism drew (and draws) on a whole range of discourses and myths in order to build its own philosophy. Alongside the growing fetish of the countryside came a nostalgia for bygone eras when people's attachment to the land and/or their attitudes to the body could fulfil the new cravings for self-actualization and spiritual plenitude. Ancient Greece provided a ready precedent for the latter, with the athleticism of its gymnasts and its body-philosophies of citizenship offering a vivid library of utopian imaginings. Gymnosophy, then, re-emerged especially powerfully in the German *Freikopekultur* and *Nacktkultur* movements.

George Mosse charts the growth of these movements in *Nationalism and Sexuality* (1985). In opposition to the decadence and degeneracy perceived as rife in the city, the German life-reformers looked to their national landscape, transplanted some ideas from Greece, and built a cult of the countryside and of the natural (national) body which embraced social nudity with fervour. Importantly, a distinction was drawn over the appropriate location for nakedness, marking a differentiation in the symbolic meaning (and reading) of the nude body: '[n]udity was acceptable only ... when seen in an unspoilt natural setting: meadows, gardens, or against the backdrop of the sea' (Mosse, 1985: 51) – in the city, or indoors, or on a stage, the naked body became lewd and sexual; in the country, it was only *natural*. As Swiss exponent of naturism Werner Zimmermann put it:

> The body of a human being who is wholly open to the sun . . . is not undressed, not truly naked. The chaste appeal of naturalness dwells on every one of its movements. Whoever walks in such a garment of light, conscious and wholly unembarrassed, is, as if by a spell, more protected from sexual desire than is the finishing-school girl with her thousand secrets, or the monk behind thick walls, or the pious sister in her massive white blinkers. (Zimmermann, 1927, quoted in Szeemann, 1997: 3)

This emphasis on naturism as non-sexual (or rather, as we shall discuss later, as sexual only within tightly prescribed limits) remains an important (if controversial) component in the discourses of the movement, played out through its own publications. The Naturist Society's website, for example, immediately stresses that its activities comprise only 'family-oriented, non-sexual nude recreation'. However, such a disavowal of the erotics of naturism did not pervade all branches of the early movements. Some naturism advocates recognized the sexual potential of being naked in nature, constructing a vision of the countryside as a 'pre-cultural' site for free love – an Eden where natural passions could flourish and where lovers could 'cast off the yoke of civilization' which might otherwise constrain their desires (Bech, 1997: 148). In the homosexual imaginary, such mythologies have long been significant (Bell and Valentine, 1995; Fone, 1983), and traces can be also found in literary, artistic and reform materials circulating in the English back-to-nature milieu. One prominent figure in this context was Edward Carpenter, who celebrated same-sex union alongside union with nature (see Tsuzuki, 1980). As he wrote in *Love's Coming of Age* (1896: 16): 'sexual embraces seldom receive the benison of Dame Nature, in whose presence alone, under the burning sun or the high canopy of the stars and surrounded by fragrant atmosphere, their meanings can be fully understood' (quoted in Marsh, 1982: 22).

Mosse (1985) makes an interesting distinction between the English and German movements which is relevant here. The homoerotic tradition he detects in *fin-de-siècle* English culture – essentially, at that time, a practice of members of the artistic and literary elites – did not focus its attention on either the nation or the national body; meanwhile, natural male beauty became a national stereotype in Germany, with male bonding and manliness central to the spirit of nationalism (though this had both socialist and fascist manifestations). However, as Matless (1995) clearly shows, by the inter-war period England's reformers were beginning to turn their attention to the national body. Moreover, the new bodily disciplines of English 'right living' had to fight to find a place within the popular imagination which was responding not to the eulogies of naked masculinities in nature penned by the likes of Carpenter, but to the bawdy humour of saucy seaside postcards and music hall smuttiness – as Matless (1995: 115) puts it, there were particularly strong 'English undertows which pull[ed] bodily exhibitionism into nudge-and-wink territory'.

As Matless concludes, '[t]wo kinds of Englishness rub against one another, one looking down on common vulgarity, and seeking to raise people from low humour to a higher body, the other popularly laughing from below at the moral heights' (1995: 115). Naturists' distaste of such outpourings of working-class bawdiness locates the movement within an essentially middle-class frame, again distinguishing its ideological underpinnings from the German mass movements at this time. Certainly, the ethics and aesthetics of English 'Sun Clubs' are marked by middle-class tastes and morals. Domesticating naturism in this way also bears the mark of a colonial impulse, revealed in the styling of clubs and clubhouses (especially when described as 'nudist colonies'), and the hierarchical, rule-bound social structures which lack the romantic-nationalist slant present in the German movement. Greta Gaard (1997) makes a link between colonialism and erotophobia that could, perhaps, draw these strands together in terms of 'civilizing nature', by imposing a set of colonial morals upon ways of being 'natural'. Such a morality still pervades the English naturist scene, especially around issues of sex.

English attitudes to naturist publications exacerbated the 'problem' of naturism's public image, especially as pornographers quickly jumped on the bandwagon (Anthony, 1998). The very first British 'sex film', according to Ian Conrich (1998), was naturist flick *Nudist Paradise*, released in 1958. More significant, perhaps, is the fact that an earlier American naturist movie, *Garden of Eden* (1954) was shown in British cinemas with a 'U' certificate (allowing even small children to watch it). British naturist films of the 1960s – such as *Naked as Nature*

Intended – carefully avoided outright sexploitation tactics (although porn films were being made under the guise of naturism at the same time), but audience reactions to such pictures – famously lampooned in the saucy British comedy *Carry on Camping* – again reveal popular English cultural attitudes to the naked body. It seems impossible to imagine a mass movement of 3 million social nudists in Britain (the current figure offered in Sam Anthony's documentary [1998] is 40,000); while one critic noted that the nudist film *Some Like It Cool* 'wouldn't bring a blush to a church outing' (quoted in Conrich, 1998: 90), it would almost certainly bring on a fit of the giggles.

Around this time other forms of social nudity were emerging, however, which were more concerned with sexual politics than innuendo. In the 1960s, nakedness as a political act became one of the expressive practices of the hippie movement – and within the doctrines of freedom (including free love), social nudity became once more squarely equated with the erotic. In communes and particularly at beaches, US hippies and radicals began to push against 'straight' social norms:

> By the mid-1960s, nudity was no longer the exclusive province of eccentric heliophiles sequestered in private clubs or restricted preserves, where they could indulge in sybaritic extravaganzas unencumbered by clothes. . . . The political left measured its radicalism by the number of square inches of skin it exposed for public consumption. Thousands of young, college-educated radicals and free spirits flaunted skin as a way of taunting adult authority figures and traditional values. Nude was natural, and natural was good. (Lencek and Bosker, 1998: 263)

Nude bathing on the beach had once, of course, been commonplace. In England, for example, complete nudity was the norm for bathers in many resorts at least until the turn of the 19th century, with the body's exposure to sea and sun seen as therapeutic rather than pornographic. Across the 19th century, however, beach nudity was progressively eroded, matched by the evolution of the bathing suit; it was not until the 1960s that naked flesh began to appear once more on the shoreline (Lencek and Bosker, 1998).

As Douglas et al. (1977) discuss, 1960s radical nudity was often enacted on 'wildcat beaches' – not designated naturist beaches, but beaches 'liberated' for illegal nude scenes. These sites were particularly associated with two groups – hippies and gay men, the latter tying beaches into the secret geography of sites for outdoor sex frequented by men (discussed later). The hippie impetus was, they later suggest, to politicize the 'liberation' of beaches for nudity and sex as part of a body politics and a politics of space which also manifested itself in communal living, 'nude-ins' and the free festival scene. Here, self-actualization means being 'free', being 'real' and being 'natural' – a trajectory we shall later trace further in contemporary embodied relations to nature, such as those of the mythopoetic men's movement. Before that, however, we want to spend a little more time considering the ways in which the natural landscape and its inhabitants are constructed as erotic, and how those erotics are utilized in the popular imagination and in sexual practices. Opening out the discussion of naked bodies in nature to incorporate eroticization gives a new dimension to the relationships we have thus far explored.

The Erotics of Nature

> Sex is our allotment – in the heart of the city, our little square of heath. To make love, we fling off our clothes, abandoning the signs which mark us as civilised: law, art and language. . . . Sex, then, has been appointed to the domain . . . within which we can experience ourselves as at once natural and human: that is, as beings continuous with the world beyond culture's indiscernible edge. (Ward, 1987: 161)

If we are to unravel a little more the complex negotiations between naturism, nature and the erotic, it might be worth considering how the rural is culturally invested with sexual meanings. As we have already noted, the countryside has long been used as a backdrop for romantic and erotic imaginings which mark rural space as pre-cultural, 'natural' – as untainted by civilization, which is here constructed as repressive (hence the particular resonance of this mythology in the lesbian and gay imaginary). Elsewhere, the erotics are more concerned with country folk, drawing on a number of stereotypes: of a rugged, rustic naivety, or of animalistic hypersexuality; of sauce or of raunch – but always as closer (sometimes too close) to nature (Bell, 2000a). From jokes about 'sheep shagging' and inbreeding to *Lady Chatterley's Lover* and *Maurice*, we see a spectrum of erotic readings of the rural and its people, which we need to register here and then feed back into our discussions of naturism – for if these discourses and imaginings work to sexualize the rural, then their traces will linger on in other ways of locating the body in nature.

In this context, a brief look at 'outdoor sex' might start this process. As Jon Ward writes, there is a particular link between sex and nature which gets articulated in fantasies and practices of sex *in* nature. And since Ward describes sex as 'our little square of heath', we will begin with the following entry from film-maker Derek Jarman's journal, evoking the erotics of London's Hampstead Heath:

> The deep silence, the cool night air, the pools of moonlight and stars, the great oaks and beeches. . . . For those in the know, the alfresco fuck is the original fuck. (Jarman, 1992: 83)

Outdoor sex scenes and sites have long been central to gay cultures, and have a special resonance which links back to the romantic imaginings of a pre-cultural (pre-homophobic) Eden sketched earlier. Gay sexual spaces emerged in Western cities often paradoxically in open space since, as George Chauncey (1996) puts it, 'privacy could only be had in public' – such spaces were (and indeed still are) often less under surveillance, less regulated, less 'visible' than either commercial or residential spaces. So, the alfresco fuck is original in the sense that it enacts a connection to and an emphasis on the role of such spaces in what Gordon Brent Ingram (1997) calls 'queerspace ecologies'.[1]

But the rural landscape is also eroticized in other ways in these scenes. Ethnographic and autobiographical accounts of outdoor sex, like Jarman's, often specifically emphasize these erotics, stressing the 'fit' between sexual practice and location:

> My best memory of outdoor sex was a bit of mutual (safe) buggery, just in boots, in the middle of the day in late summer, with the sun glinting through the leaves of the forest. We ran naked through the trees, rubbing against the bark, and rolled in the leaves ... ('Jon', in National Lesbian and Gay Survey, 1993: 116)

> Woodland in the dark has a promise of adventure. Perhaps it is the combination of fear and expectation. Behind each tree you may expect something or someone frightening or tempting or both. For me it is very erotic scenery ... leathersex and a scenery like these woods belong to each other. ('Richard', in van Lieshout, 1997: 354)

Of course, outdoor sex is not exclusively a queer pastime; the descriptions and motives of heterosexual participants share much in common with those cited above. In *The Nude Beach*, Douglas et al. lead into their discussion of nudism via a consideration of outdoor sex in America:

> ... a large percentage of adults have made love under the stars at some time. Lots of people simply make love in the back yard on a summer evening after the kids have gone to sleep. Others seek out a lonely mountain path by a stream, but even more find the beaches an especially beautiful and romantic place. They find a sense of freedom and beauty in the sound of the waves, the smell of the salt, the feel of the soft sand under their bodies, and the feeling of the breeze on their bare flesh. (1977: 30)

Here we find distilled the 'special feel' of natural settings articulated throughout the overlapping experiences of naturism and outdoor sex – only here it is eroticized, whereas the discourses coming from the contemporary naturist movements continue to disavow any erotics. So, while Douglas et al. talk of 'a certain delightful sense of naughtiness and abandon' found in 'being one with nature' (1977: 32), such potentialities are denied by naturists' stress on 'non-sexual nude recreation'. That last word – *recreation* – is an especially potent one. The practices of naturism and outdoor sex we are dealing with here are essentially metropolitan modes of escapism: the countryside offers recreation and regeneration for city folk. For those born and raised in rural areas, where outdoor sex is more of a desperate necessity than an opportunity for delightful naughtiness, such an erotics can be entirely absent, or at least framed in radically different, non-metropolitan ways (Bell, 2000b; Kramer, 1995; Phillips et al., 2000). As with the romantic longings of the back-to-the-land milieu, the landscape is often essentialized and idealized in a way that works to deny some of the raw realities of country life – although a puritanical ethos of simple, authentic 'country living' continues to pervade some Sun Clubs, other naturist sites have all the facilities required by modern metropolitan lifestyles. Whatever the practicalities, however, the countryside maintains a huge cultural appeal based on its symbolic power in re-energizing urban dwellers – an appeal manifest in the many forms of recreational escapism practised there, by naturists and 'textiles' (clothed people) alike.

The Call of the Wild

Constructions of the rural landscape as a regenerative resource also find echoes through the narratives of the mythopoetic men's movement, which has its own take on nature's benefits to the bodies (and minds and spirits) of city men. The inner workings, dominant discourses and political orientations of the movement have been ably and critically covered elsewhere, and need not overly detain us here (see Bonnett, 1996; Messner, 1997; Pfeil, 1995). In sketch form, the movement's key concerns lie in rediscovering and celebrating 'deep masculinity' as a response to the softening effects of feminism, white-collar alienation and postmodern urban living; tapping historically and transculturally into sources of masculine strength; reaffirming via naturalization the duality of the sexes, and the social roles attached to each side of that duality; and – and this is where our primary interest lies – *locating* a space for rituals of deep masculinity: in the wilderness. Touching on the 'wilderness philosophies' of writers like Henry David Thoreau and drawing on Native American mythologies and rituals (alongside a grab-bag of 'primitive' resources which also includes African tribal custom, Old Norse myths and Greek mythology), the movement looks to the wilderness as a pre-industrial and non-westernized space of retreat and remasculinization:

> Enter the wilderness.... The Earth Father welcomes us, challenging us to become stronger and deeper as men. We are at home in nature. (Kipnis, 1992: 163; quoted in Bonnett, 1996: 283)

Crucially, in this wilderness gender relations can re-emerge in their 'natural' form, and men can once again become men. Becoming a man, then, is a process enacted in close, communal contact with nature, through ritual practices (initiation rites, drumming, sweat lodges). In a similar way, Rachel Woodward (1998: 287) discusses the role of the rural in contemporary British soldier training, for making a particular kind of man through no less ritualized practices: 'This man, embodying some of the qualities associated with hegemonic masculinity in contemporary Anglo-American culture, is physically fit, mentally brave, emotionally hard' – the main difference being in relation to the last phrase, since the mythopoetic men's movement stresses forms of emotional work (dealing with 'wounds', bonding rites, talking circles) that are certainly out of kilter with military training. As Woodward notes, the military gaze on the landscape is emotionless, rational, cartographic; any glimmer of emotional or aesthetic appreciation (other than, perhaps, that cloaked in nationalism) is quickly suppressed as 'unmanly'.

In another important way, however, military and mythopoetic masculinities have a common agenda around 'manliness': both work strenuously to maintain a homosocial-heterosexual proscription. As the 'gays in the military' debate in the USA has all too clearly shown, the 'threat' of homosexuality is constructed as omnipresent but constantly, phobically policed in such contexts (Butler, 1997). The cultural construction of homosexuals as hypersexual, exposed in the logic of the US military ban, obviously figures equally in the environment of the mythopoetic men's movement (see Collier, 1996), just as it did in the early naturist movements – and, indeed, in the (more or less explicit) prohibitions of today's naturists. Charged with contaminating and compromising the safety of the homosocial-heterosexual matrix, lesbians, gay men and bisexuals are dismissed as 'perverts' on the 'family-oriented' naturist scene just as they are denied a voice within the mythopoetic men's movement.[2] As a 'moral panic' within any of these contexts, homosexuality exposes the central problematic of these bodies in nature: sex.

The Sex of Nature and the Nature of Sex

English naturism has had to weather a number of such moral panics, all of which centre on what kinds of sex fit in with the naturist reading of nature. Put bluntly by Ralph Rugoff (1995: 181), '[n]aturists don't fuck, they breed'. Along with homosexuality, then, UK naturists have had to deal with the 'threat' of permissiveness and promiscuity (most notably in the 1960s) and the 'spectre' of paedophilia, as well as the slippage between naturist publishing and pornography (Anthony, 1998). The (perhaps peculiarly English) way of dealing with this has been to tighten the rules and regulations of naturist clubs and societies: no singles/families only, no 'sexual body jewellery', no photographs, etc. Expansive public relations work has also shored up the 'respectable' image of naturism (to the annoyance of some factions), leading to its effective *privatization* into coyly named Sun Clubs and isolated, rigidly demarcated stretches of beach surrounded by warning signs.

Clearly, procreative, familial, adult, heterosexual intercourse is the only form of sex to be equated with nature here; even then, it has to be hidden away – outdoor sex is bad PR, especially while the National Trust, or the press, or some generalized notion of the (offended or offendable) 'public' might be watching.[3] The perceived social offence of public nudity, which rests on its sexual associations, has its spatial manifestation in segregated, privatized sites – unless, as in the case of wildcat beaches or protest-oriented nakedness (streaking, for instance), social offence is an active motivator for disrobing. The discourse of contemporary naturism, of course, stresses that only textiles find the naked body erotic; the normalization of nudity effectively desexualizes it.

Moreover, any forms of sex other than procreative, familial, adult, heterosexual intercourse are clearly constructed as 'against nature'; or, perhaps, they get relocated into the culture (or civilization) which is cast off when the body is unadorned (to recall Jon Ward). In this logic the body is re-naturalized and simultaneously de-sexualized; rather than casting off culture/civilization to release either animalistic or 'natural' (culturally suppressed) passions – a viewpoint we see in some of the lesbian and gay imaginings of a 'queer Eden' – a strict code of morality is instead *naturalized*. In the same way, the mythopoetic men's movement seeks to 'displace gender issues from social or economic consideration onto the mythical terrain of Nature' (Bonnett, 1996: 288). Nature is thus simultaneously civilizing and the other of civilization, and it is this paradox which the bodies in nature we have been discussing articulate.

The complex issues around naked bodies in nature which we have begun to discuss here clearly warrant further exploration. More comparative readings of the histories of naturism in different national contexts would no doubt reveal the web of ideological and philosophical undergirdings which have resulted in distinct attitudes to nakedess and to nature in different places. The sketches of German and British naturism presented here suggest as much. In addition, it would be fruitful to explore more fully the different constructions of nature articulated by different naturist practices: Sun Clubs are obviously very different spaces from wildcat beaches, for example. The relationship between nature, nudity and the erotic has been shown here to be particularly complex – while there are structuring logics at work within naturist discourses which work to disavow the erotics of nature, evidence from accounts of sexual encounters suggests a powerful sexualizing impulse associated with nature in other contexts.

Further examination of the interplay between nature and sex might begin to establish the ways in which nature works as an erotic construct. And, via the brief discussion of the mythopoetic men's movement, we have suggested that nature can be mobilized in order to 'naturalize' gender as well as sexual relations (seen also in some aspects of ecofeminist discourse; see Mies and Shiva, 1993). Clearly, then, in many different ways, nature is constructed as exerting powerful influences upon the human bodies in its midst; the removal of our clothes, as we have seen, connects us to particular imaginings of nature and of what it means to be 'natural'.

Notes

Thanks to John Urry, Phil Macnaghten and three anonymous referees for their helpful comments. Thanks to Chris Thomas for *The Beach*. Ruth Holliday's research for this article was supported by a grant from the ESRC (R000236657).

1. It is similarly 'original' in its reference to the Garden of Eden and to original sin, of course.
2. We should not forget the wilderness-based queer practices of either the 'radical fairies' or rural lesbian separatists, where the links between nature and the body work very differently (see Gaard, 1997; Thompson, 1987; Valentine, 1997).
3. For an interesting discussion of the latter logic in relation to indecent exposure – 'flashing' – see Kirby (1996).

References

Anthony, Sam (dir.) (1998) *Witness* (TV documentary about British naturism). London: Channel 4 Television (broadcast 22 November).

Bech, Henning (1997) *When Men Meet: Homosexuality and Modernity*. Cambridge: Polity Press.

Bell, David (2000a) 'Eroticizing the Rural' pp. 83–101, in Richard Phillips, David Shuttleton and Diane Watt (eds) *Decentring Sexualities*. London: Routledge.

Bell, David (2000b) 'Rurality, Masculinity, and Homosexuality', *Rural Sociology* (in press).

Bell, David and Gill Valentine (1995) 'Queer Country: Rural Lesbian and Gay Lives', *Journal of Rural Studies* 11(2): 113–22.

Bonnett, Alastair (1996) 'The New Primitives: Landscape and Cultural Appropriation in the Mythopoetic Men's Movement', *Antipode* 28(3): 273–91.

Bunce, Michael (1994) *The Countryside Ideal: Anglo-American Images of Landscape*. London: Routledge.

Butler, Judith (1997) *Excitable Speech: A Politics of the Performative*. New York: Routledge.

Carpenter, Edward (1896) *Love's Coming of Age*. London: Hutchinson.

Chauncey, George (1996) '"Privacy Could Only be Had in Public": Gay Uses of the Streets', pp. 224–61 in Joel Sanders (ed.) *Stud: Architectures of Masculinity*. New York: Princeton Architectural Press.

Collier, Richard (1996) '"Coming Together"? Post-Heterosexuality, Masculine Crisis and the New Men's Movement', *Feminist Legal Studies* 4(1): 3–48.

Conrich, Ian (1998) 'Forgotten Cinema: The British Style of Sexploitation', *Journal of Popular British Cinema* 1(1): 87–100.

Douglas, Jack, Paul Rasmussen and Carol Flanagan (1977) *The Nude Beach*. Beverley Hills, CA: Sage.

Fone, Byrne (1983) 'This Other Eden: Arcadia and the Homosexual Imagination', *Journal of Homosexuality* 8(1): 13–34.

Gaard, Greta (1997) 'Toward a Queer Ecofeminism', *Hypatia* 12(1): 114–37.

Ingram, Gordon Brent (1997) '"Open" Space as Strategic Queer Sites', pp. 95–125 in Gordon Brent Ingram, Anne-Marie Bouthillette and Yolanda Retter (eds) *Queers in Space: Communities/Public Places/Sites of Resistance*. Seattle, WA: Bay Press.

Jarman, Derek (1992) *Modern Nature: The Journals of Derek Jarman*. London: Vintage.

Joad, Cyril (1934) *A Charter for Ramblers*. London: Hutchinson.

Kipnis, Aaron (1992) 'In Quest of Archetypal Masculinity', pp. 152–5 in C. Harding (ed.) *Wingspan*. New York: St Martin's Press.

Kirby, Kathleen (1996) *Indifferent Boundaries: Spatial Concepts of Subjectivity*. New York: Guilford.

Kramer, Jerry-Lee (1995) 'Bachelor Farmers and Spinsters: Gay and Lesbian Identities and Communities in Rural North Dakota', pp. 200–13 in David Bell and Gill Valentine (eds) *Mapping Desire: Geographies of Sexualities*. London: Routledge.

Lencek, Lena and Gideon Bosker (1998) *The Beach: The History of Paradise on Earth*. London: Secker and Warburg.

Marsh, Jan (1982) *Back to the Land: The Pastoral Impulse in Victorian England from 1880 to 1914*. London: Quartet.

Matless, David (1995) '"The Art of Right Living": Landscape and Citizenship, 1918–39', pp. 93–122 in Steve Pile and Nigel Thrift (eds) *Mapping the Subject: Geographies of Cultural Transformation*. London: Routledge.

Matless, David (1998) *Landscape and Englishness*. London: Reaktion.

Messner, Michael (1997) *Politics of Masculinities: Men in Movements*. Thousand Oaks, CA: Sage.

Mies, Maria and Vandana Shiva (1993) *Ecofeminism*. London: Zed Books.

Mosse, George (1985) *Nationalism and Sexuality: Middle-Class Morality and Sexual Norms in Modern Europe*. Madison: University of Wisconsin Press.

National Lesbian and Gay Survey (1993) *Proust, Michelangelo, Cole Porter, Marc Almond and Me: Writings by Gay Men on their Lives and Lifestyles*. London: Routledge.

Pfeil, Fred (1995) *White Guys: Studies in Postmodern Domination and Difference*. London: Verso.

Phillips, Richard, David Shuttleton and Diane Watt (eds) (2000) *Nonmetropolitan Sexualities*. London: Routledge.

Rugoff, Ralph (1995) *Circus Americanus*. London: Verso.

Szeemann, Harald (1997) 'Here I Am Human, Here I Am Free', pp. 1–5 in Michael von Graffenried, Harald Szeemann and A.D. Coleman, *Naked in Paradise*. Stockport: Dewi Lewis.

Thompson, Mark (1987) *Gay Spirit: Myth and Meaning*. New York: St Martin's Press.

Tsuzuki, Chushichi (1980) *Edward Carpenter 1844–1929: Prophet of Human Fellowship*. Cambridge: Cambridge University Press.

Valentine, Gill (1997) 'Making Space: Lesbian Separatist Communities in the United States', pp. 109–22 in Paul Cloke and Jo Little (eds) *Contested Countryside Cultures: Otherness, Marginalisation and Rurality*. London: Routledge.

van Lieshout, Maurice (1997) 'Leather Nights in the Woods: Locating Male Homosexuality and Sadomasochism in a Dutch Highway Rest Area', pp. 339–56 in Gordon Brent Ingram, Anne-Marie Bouthillette and Yolanda Retter (eds) *Queers in Space: Communities/Public Places/Sites of Resistance*. Seattle, WA: Bay Press.

Ward, Jon (1987) 'The Nature of Heterosexuality', pp. 145–69 in Gillian Hanscombe and Martin Humphries (eds) *Heterosexuality*. London: Gay Men's Press.

Woodward, Rachel (1998) '"It's a Man's Life": Soldiers, Masculinity and the Countryside', *Gender, Place and Culture* 5(3): 277–300.

David Bell teaches Cultural Studies at Staffordshire University. He is co-editor of *Mapping Desire* (1995), and co-author of *Consuming Geographies* (1997) and *The Sexual Citizen* (2000).

Ruth Holliday teaches Cultural Studies at Staffordshire University. She is co-editor of *Organization/Representation* (1998) and *Contested Bodies* (2001), and co-author of *Pleasure Zones* (2001).

Action and Noise Over a Hundred Years: The Making of a Nature Region

DAVID MATLESS

Conduct: The Cultural Grain of a Region

Moral Geographies

> M. Keane. Fined £155.00 with £45.00 costs at the Lowestoft Magistrates' Court on the 8 October 1993 for navigating a hired motor cruiser at a speed in excess of 4 mph, navigating without care and caution, failing to conform with instructions of the River Inspector and using obscene, scandalous, abusive, indecent or improper language on the River Waveney at Beccles on 11 April 1993. (entry in 'Skipper's Handbook' on hired cruiser, July 1995)

Hiring a motor cruiser on the Norfolk Broads in July 1995, and dutifully reading the 'Skipper's Handbook', I came across the story of M. Keane, a warning of what could happen if, at any point over the following week, control of boat or temper was lost. After instructions on engines and mooring and the disposal of wastes, the handbook gave cautionary tales of what not to do on the Norfolk Broads. This was more than just a matter of bye-laws. From its moment of 'discovery' in the late 19th century (McWilliam and Sekules, 1986; Taylor, 1995) this region of shallow lakes and rivers in eastern England has been defined through certain modes of conduct. Keane's swearing and speeding incident is in part a story of illegality, a warning to the visitor that in the last resort the courts will function, but is also a tale of modes of being.

Keane's story, and this article, concern nature and the social/anti-social body. The making of the Broads as a nature region – the dominant definition of the place today – cannot be understood apart from the interweaving of conduct and environment in various moral geographies (Driver, 1988; Matless, 1994), with

Body & Society © 2000 SAGE Publications (London, Thousand Oaks and New Delhi),
Vol. 6(3–4): 141–165
[1357–034X(200009/12)6:3–4;141–165;015475]

modes of bodily conduct and their related technologies defined as in or out of place (Cresswell, 1996; Stallybrass and White, 1986) by those claiming the cultural authority to define the nature of the region. Such moral geographies are integral to the regional substance. Care for nature, joy in environment, laborious drudgery, local identity, the making of money, all come together in and are woven through the conduct of bodies in landscape.

'Nature' is a key defining and legitimating element in the region's contested social geography. The analysis of nature and society here does not begin by defining the two as distinct, with the social acting on or taking place within the natural. Rather, that which is valued as nature is produced through the social spaces for which it provides constituent elements; earth, air, water, vegetation, etc. Life in this region turns on different ways of dealing with such elements, dealings which combine to render a place of complex and variegated geoaesthetics (Matless, 1997a). The approach taken here extends Nikolas Rose's (1996) Foucauldian arguments on governmentality, corporeality and spaces of subjectification to consider the production of the bodily subject through landscape, where that landscape is not simply a backdrop to action but a culturally charged object (for the sense of 'landscape' see Matless, 1998). Landscape becomes the subject of codes of conduct and aesthetics of existence (Foucault, 1986; Thacker, 1993). The region is put in question, and thereby produced as a defined region, through what Rose terms a 'spatialization of being', and through this regional spatialization the body gathers a becoming or unbecoming sociality:

> . . . repertoires of conduct are activated that are not bounded by the enclosure formed by the human skin or carried in a stable form in the interior of an individual: they are rather webs of tension across a space that accord human beings capacities and powers to the extent that they catch them up in hybrid assemblages of knowledges, instruments, vocabularies, systems of judgement and technical artefacts. (Rose, 1996: 143–4).

This article addresses the social body by playing off different repertoires of conduct which have made up the cultural grain of the region over a hundred years, repertoires which have cut across and against one another with effects of friction and frisson.[1]

The geography of this story is not only internal to the region, in part of course as no region, not least one driven by tourism, makes sense outside its relations to other places. The individual region, however, is also defined relationally as a spatial *type*, and thereby cast into debates concerning water and wetland as particular forms of social space (see Bachelard, 1983; Cameron, 1997, 1999; Corbin, 1994; Giblett, 1996; Goubert, 1989; Matless, 1992; Wright, 1999). Two cultural formations come into play here. The first is the *wetland* as a zone of ecological value and boggy, intimate and carefully observant contact between humans and nature. This version of waterland attends as much to the land as to the water, or rather presents a space defined by the intermingling of the two such that categories of liquid and solid are blurred. The second is the *water* as a space of play, a zone akin to the seaside where water and land are clearly demarcated by riverbanks or lakesides, and certain kinds of fun take place afloat and ashore. The Broads make sense in terms of these two versions of liquid social space, which bind self and environment in very different ways such that tensions of conduct in the region relate in part to the different human ecologies of rivers, lakes, reedbeds, marsh, carr woodland. Cultural territories are staked out over different riparian spaces, and zonal culture clashes proceed (Matless, 1994: 145–7). The region also refracts national and global transformations in the culture of water since the late 19th century concerning pollution, patriotism, tradition, purity. Such changes and conflicts do not simply indicate different cultural readings of a constant underlying nature; rather, the status of water and wetland shifts according to what Illich terms 'the historicity of "stuff"' (1986: 3–4). Different configurations of water act as the 'ground' for various forms of identity, the moral geographies of the region working in part through its exemplification of a valued landscape type.

Moral Histories

This article forms part of a wider project on cultures of landscape in Broadland over the past century. The aim is not to tell a comprehensive historical story, but rather to attend to the active presence of historical narratives within the region in a way which might unsettle current orthodoxies. The strategy here is in part to bring displaced pasts back into the regional story. Processes of cultural displacement work not only through moral geographies but moral histories, with regional history told such that some forms of past conduct become fine tradition while others appear as deviations from the regional grain. Stories of the past weave through the actions of bodies in the present.

Tom Williamson's recent landscape history (1997) presents the Broads as a historical industrial landscape in a way which sits uneasily with contemporary formulations of the region as essentially a nature zone. This article seeks a similar effect in relation to elements of leisure culture; certain forms of popular leisure have been written out of regional history as regrettable, polluting actions which have left only a stain on the place. Efforts to make Broadland a popular nature space downplay another form of popular culture. Such historic displacement is clear in the work of the Broads Authority, the conservationist statutory authority in the region, which was given National Park status in 1988 (Broads Authority, 1993; Ewans, 1992; George, 1992). Promoting 'quiet enjoyment' and styling the Broads as a beautiful and mysterious 'Last Enchanted Land' rescued only by recent conservation action, Broads Authority Chief Executive Aitken Clark tells a story of a post-war fall from grace:

> If on a dank and misty afternoon in autumn in the late 1970s, you had driven from Norwich along the 'Acle Straight', the road used by thousands of holidaymakers heading through the Broads towards Great Yarmouth, you might have seen little romance in the mist, and nothing but desolation in the flatness and in the derelict windpumps. If you had travelled to Wroxham and taken a diesel-powered launch along the River Bure, you might have wondered where you were supposed to find peace and tranquility [*sic*].
>
> Indeed, in the sixties and seventies, the candyfloss and seaside atmosphere of chip shops, amusement arcades and souvenir shops selling trinkets and kiss-me-quick hats seemed suddenly to outweigh the quiet reedy reaches, where bitterns boomed unseen and the sails of a lone yacht drifted across the horizon. You could have been forgiven for believing the scare stories and blaming the heavy hand of commercialism that had taken its toll, exploited and spoilt for ever this fragile enchanting land. Once perhaps it may have made the grade as a national park – but alas, no longer.
>
> Perhaps it was too late for the Broads. (Clark, 1989)

This passage comes from Clark's introduction to a photographic volume by Richard Denyer, who has produced a highly pluralist aesthetic of Broadland life and landscape (Denyer, 1989); Clark's piece is followed by a critical essay on 'representation and reality' by historian Sarah Knights (1989). If Clark's demonization of the recent past goes against its immediate critical and pluralist context, however, his reading of past and present indicates the dominant narrative in the Authority's work. Clark mobilizes key elements of regional iconography. Derelict windpumps stand for a fall of traditional land management. The bittern, symbolic bird of the region, with its strange booming call, is set against the sights and sound of mass leisure. The seaside resort of Great Yarmouth, at the mouth of the Broadland rivers, haunts the region; its holidaymakers rushing through the landscape along the straight Acle to Yarmouth road, the chief route for tourists from the Midlands and North, and its values migrating up-river to bring candyfloss and chips and arcades. The day launches are diesel powered. The candyfloss is sticky. Kiss-me-quick hats signify the wrong kind of romantic landscape. Recent regional history is defined as going against the regional grain, as essentially not regional.

Moral histories, then, run alongside moral geographies, indeed are always co-constituted as geohistories, with the making of the Broads as a nature region turning on the nature of Broadland as social space. A final introductory point should be made here on sources. This article deals primarily with written and pictorial sources alongside field observations, treating the former as active interventions in the region which have circulated through it and beyond as a part of its production. The article is emphatically not concerned with 'images of Broadland', and would not follow any ontological or epistemological distinction of the imaginative and the practical. I allude here to Catherine Nash's recent criticisms of writings on 'nonrepresentational' theories of performative practice which, Nash suggests, have argued against dualisms of representation and practice through a rhetoric which paradoxically upholds the latter against the former and thereby remains caught in the distinction (Nash, 2000).[2] This article is concerned with practices of writing and picturing which have gone towards making up the region, and through which we can move into broader matters of what the region has been, is and might be.

Pleasure in the Balance

The remainder of this article first considers the production of Broadland as a site for improving pleasures predicated on a balance of mind and body, before moving on to the brash fun considered by the Broads Authority to be out of place. Modes of enjoyment are placed in the balance. The article concludes by addressing other formulations of disciplinary regulation or bodily freedom.

Space of Improving Fun

One version of Broadland is of the region as a space for the improvement of the self through the experience of nature, whether in the form of educational observation, spiritual communion or physical exercise. Today the Broads Authority seeks to harness programmes of environmental education to the restoration/improvement of the landscape, such that a previous assumed downward spiral of bad environmental management and bad environmental experience can be reversed for the improvement of both place and people. This particular social aesthetic has structured commentary on Broadland since the region was defined as a leisure and/or nature space.

In October 1893 the illustrated London periodical *The Graphic* recounted a story of 'A Week In A Wherry' (Figure 1). The University College Hospital Working Lads' Institute had organized a week on the River Bure for a party of 40 East End boys on board three wherries; many such former cargo vessels were then being converted for leisure hire. Such boats were more commonly hired by upper middle-class visitors who would rent a local crew along with the boat; the urban working-class presence in Broadland was generally restricted to those taking day-trips on large steam vessels from Great Yarmouth. The presence of urban working-class boys, supervised by eight doctors and students, is seen to be appropriate here because of the improving nature of the exercise, which follows the pattern of much open-air activity organized by youth and philanthropic organizations at this time across the political spectrum (Rosenthal, 1986; Springhall, 1977;

Figure 1 East-End boys in Broadland (from *The Graphic*, 28 October 1893: 544)

Taylor, 1997). Physical health and moral improvement are to be achieved through the release of pent-up young male energy and the simultaneous inhalation of air and exposure to light. Exertion makes the body function anew in terms of both a letting go and a taking in.

It is unclear as to whether *The Graphic*'s illustrator G. Durand was present on the trip; the likelihood is that these pictures were in part made up of stock drawings, such as that of the boys at table, which were then superimposed on a generalized flat river landscape. Michael Wolff and Celina Fox (1973) note that *The Graphic* still used such stock blocks in the 1890s. Durand's pictures tally with incidents recounted in the text, such as the 'combined athletic sports' for visitors and villagers at Horsey. The boys learn to swim, sail, clean and cook in a spirit of self-reliance in basic conditions. The doctors and students stress to them the Englishness of water: 'to teach them that English boys should regard it as their home equally with the land'. Waterland becomes patriotic space. The whole trip amounts to:

> ... a branch of prophylactic medicine; and, as each wherry carried a qualified doctor on board, they can testify to the increased health and vigour carried back by their youthful charge, as well as the social benefits and large vistas of life opened up. (*The Graphic*, 1893: 545)

The Graphic's story of social improvement through country/nature experience is echoed in the work of figures such as George Christopher Davies, credited with popularizing the area in the late 19th century. Davies promoted the Broads as both leisure space and nature region, the two roles mediated through appropriate conduct benefiting both place and person. Davies's 1876 boys' adventure story *The Swan and Her Crew* provides an upper middle-class variant of *The Graphic*'s account, pitting sporting boys who make their masculinity through combining shooting and nature conservation against others who destroy nature in less focused fashion. The good sportsman aligns killing and conservation, the non-sportsman kills regardless, and the former gains improvement in mind and body (Mackenzie, 1988). As in *The Graphic*, the experience of a nature region gives health, transforming the sickly and bookish Dick Merrivale from a pale, delicate aesthete into a sturdy lover of beauty: 'I did not know that life could possibly be so jolly, until I learnt something of natural history' (Davies, 1932: 114). Health enables Dick and his friends to engage in class warfare for the Broads-as-nature-region, beating up Yarmouth 'street-boys' who are smashing eggs for fun. If in *The Swan* Davies dramatized nature conduct for a popular middle-class boy readership, in his guidebooks he specified rules of conduct for the general visitor to the Broads. In his *Handbook to the Rivers and Broads of Norfolk and Suffolk*, published in 1882 and into 50 editions by 1923, Davies introduced the region by stating some 'hitherto unwritten rules', written in a light-heartedly serious manner to deter visitors from lighting fires, speeding, flower-picking and noise: 'Don't play the piano in season and out of season (the reedbird's song is sweeter on the Broads)' (Davies, 1923: 13).

Similar schemes of educational and behavioural value run through later work aimed at children and adults. If Davies's late Victorian stress on manly shooting declines – though *The Swan* is republished as late as 1932 – oppositions setting good conduct on the side of regional nature and bad conduct against it are strengthened, with the ideal mode of being in Broadland entailing alert observation, self-reliance and care for nature. The piano's role as disturber of the peace is usurped by the gramophone, and later the transistor radio. In *Coot Club* (1969) and *The Big Six* (1940) Arthur Ransome provided children with adventure stories 'symbolic of the forces contesting the future of the Broads' (Brogan, 1984: 337). Ransome holidayed on the Broads throughout the 1930s, and *Coot Club*, especially, presents an environmental morality tale where the adventuring children of the professional class ally with good locals against bad locals and bad visitors to defend nature against the wrong kind of culture. Dick and Dorothea Callum, the visiting offspring of an archaeologist, join with Tom Dudgeon, son of a Horning doctor, to act as vigilantes for the local wildlife. Tom is in a Bird Protection Society, and defends nesting coots against thoughtless trippers who speed and make waves. Such people, 'Real Hullabaloos' (Ransome, 1969: 42), perform like characters from a later Public Information Film on how not to behave on the river, with the chief villain of the piece being a large cruiser, the *Margoletta*, carrying a mixed company of fashionably blithe trippers fond of playing a gramophone on deck and ignorant of the nature of the region. Tom sets them adrift when they moor by a coot's nest. The tripper here, as in much later commentary, appears as a kind of polluting pond-skater, slipping across the surface of local ecology and having no meaningful connection to it, but managing to destroy it in the process. A culture of nature is defined against another way of being, though it should be stressed that this is not simply a local/non-local story. George Owden, a local working-class boy in league with the Hullabaloos, betrays the children and the coots in both of Ransome's Broads tales, while Ransome's particular social democratic philosophy of improvement through nature demands that the Coot Club itself is a cross-class alliance including locals and visitors. Neither the forces of good nor those of evil are specific to one class or locale. The Hullabaloos come to grief in the end, holed after ramming a post in the middle of the expansive Breydon Water, and Ransome stages a local version of the *Titanic* disaster, with the gramophone 'pouring out its horrible song' as the boat begins to sink (Ransome, 1969: 327). The Hullabaloos are rescued, and humbled, by the 'Death and Glories', three boat-builder's sons from Horning.

Technologies of movement are a key part of this moral geographic tale. The children are keen sailors, and Ransome sets the mental and physical activity of sailing against the inactivity of motor cruising. Laurier (1999) has recently discussed 'classic' yachting in terms of the reproduction of social elites and the corporeal 'boundary work' involved in maintaining elite social space (see also Laurier, 1998). While the term 'elite' does not capture the social complexities in Ransome's work and in Broadland sailing more generally, an analogous process can be traced concerning the bodily dispositions and competences making up moral geographic boundaries on the water. Sailing minds and bodies are held to be alert, dextrous, coordinated within themselves and between each other. The technology of the sailing boat is seen to demand individual and social coordination enacting a complex set of practical knowledges, whereas the motor cruiser, given basic competences of steering and mooring, simply allows (rather than demands) a lazy individual demeanour. When the individual is incompetent even here, this simply underlines their lack of coordination, and when an individual (usually male) becomes over-committed to the motorized machine, acting the captain in a peaked cap as if on the bridge at sea, it is taken as evidence of a ridiculous masculinity lacking a proper sense of technology. It is therefore not only that sailing is held to tone mind and body while cruising is not, but that sailing technology demands a tighter alliance of boat and human. The hands become a focal point of sailing energy, with experienced boat handling leaving its approving mark. In *Coot Club* Dorothea wishes that others could see the transformation of her and Dick's hands: 'They are so beautifully horny' (Ransome, 1969: 178). Ransome pokes a little fun here at Dick and Dorothea's sailing aspirations, going on to describe them sailing the boat 'without a hand on the tiller other than the beautifully horny ones' (1969: 178); the satire here is in part at Dorothea's wish to *display* hands to others, rather than simply use them. The overall message, though, is that if Dorothea's language and manner remain a little refined, in keeping with Ransome's tendency to send her into self-consciously feminine fantasies on the natural world, the children's bodies are being grounded in a way which will also transform them intellectually, making them alert to the place, its wildlife, its traditions, its people.

There is not space here to trace the development of these cultures of boating from the 1930s to the present. Distinctions of craft and how they are handled remain important today, with boat hire companies re-branding themselves from an image of indulgent luxury cruising towards non-polluting motor craft, the reduction of wash through the re-design of hulls, and the promotion of electric boats, the latter presented as having been common on the Edwardian Thames. Nature, tradition and technology are enfolded as visitors are invited to tour the interiors of Broads Authority nature reserves on board the *Electric Eel*. Ransome's work, too, has been revived in the contemporary landscape. Where only a few years ago Ransome's stories had no presence even in Broads Authority information centres, now *Coot Club* and *The Big Six*, along with his Lake District stories, are for sale, alongside other water works such as Kenneth Grahame's *The Wind in the Willows*, set in the Thames Valley. A tradition of classic water literature, with its resonant social and aesthetic commentary, is aligned with the vision of the Broads as enchanted nature region, paralleling the use of literary traditions in other national park areas, notably the Lake District (Macnaghten and Urry, 1998; Urry, 1995). Aitken Clark makes the connection when interweaving ecological and behavioural degradation in his story of the post-war decades:

> It was against this background that the Broads Authority was set up in the late 1970s. Many people who remembered the idyll of Arthur Ransome, the crystal clear water, the days when sails outnumbered motor boats and when the bittern's boom was more often heard than that of the ghetto blaster were sceptical. The task facing the Authority was certainly a challenging one. (Clark, 1989)

One could argue that the Authority's strategy has been in effect a revival of the social aesthetic of *Coot Club*. The late 1980s also saw Ransome re-emerge through Roger Wardale's (1988) popular guide to *Arthur Ransome's East Anglia*; the story here parallels that told by Mike Crang in an urban context of old photographs, postcards and local memories, whereby seemingly lost ephemera of history are re-collected to produce particular popular senses of the local past (Crang, 1996; Samuel, 1994). Wardale's act of cultural recovery is aided by Ransome's standing as a writer with an acute sense of place, devoted to topographic and cartographic accuracy, and carries a moral geographic commentary on old and new days which echoes Ransome's own scheme of value. Wardale is happy to find that in some places things haven't changed too much, but laments the increased number of motor cruisers, 'chromium plated plastic-tubs' very different from 'traditional Broads craft' (Wardale, 1988: 21). To find 'the world of the Coots', Wardale says, 'it is best to visit the area out of the holiday season' (1988: 8). One could of course argue that to find Ransome Country one should visit in *peak* season, where one might relive the action of *Coot Club* and sabotage some revellers. Or perhaps the Broads Authority's acts of cultural restoration have effectively seen off the Hullabaloos.

On Not Hearing the Voice of Former Holidaymakers

> Codes of law and morals, or religious systems, never have much room in them for a humorous view of life. (Orwell, 1961: 176)

Browsing in the local history sections of bookshops, or in the information centres set up by the Broads Authority, it is difficult to catch the historical voice of the happy holidaymaker. Motivations for visiting Broadland are of course many and varied, and one should not assume that all hirers of motor cruisers in the past or present have been out for the kind of fun demonized in the accounts discussed above. If not all visitors are Hullabaloos, however, there are other variants of fun beyond the improving kind which have undoubtedly had a significant presence. And Hullabaloo-like activity is as much part of a culture of landscape as bird-watching, sailing, etc. It is difficult, however, to trace such values in the conventional history of the region as presently told. Here we explore some ways into this different cultural history, and the reasons why this mode of experiencing Broadland, shared we may assume by thousands of people over several decades, figures only in a negative fashion in current stories.

What might be termed 'non-improving fun' has both a crucial presence and absence in the contemporary and historical landscape. On the one hand it is present as an other to cement a culture of nature, to reassure it of its own propriety, on the other it is absent from valued history, appearing only in stories of decline. In the narrative of post-war environmental decline, the growth of boat hiring becomes grouped with run-off from agricultural land and sewage outflow into the rivers in a story of declining water quality. Pollution and churning of water from cruisers is now generally considered a minor factor in the process of eutrophication, whereby water has become cloudy with the loss of plant and animal species (George, 1992), but history is presented in such a way as to group all such forces in a story of environmental decline through misguided progress. Certain forms of boat conduct are defined as being against the water. There is little recognition that if one considers conservationist accounts from the immediate post-war period through to the Nature Conservancy's 1965 *Report on Broadland*, a dominant conservationist concern is over the disappearance of the Broads through vegetation encroachment, and a crucial factor in maintaining open water is seen to be boat hire. The motor cruiser could be presented as a beneficent ecological force.

Narratives of post-war environmental decline and subsequent Broads Authority-led restoration therefore write over some ecological complexities. The effect is also to historically erase the pleasures of thousands in what Nicholson-Lord has termed the 'high summer of industrial society' (Nicholson-Lord, 1987, quoted in Wilson, 1992: 14). Nicholson-Lord uses the term sardonically to denote unreflective consumption and environmental exploitation, but the phrase might also be used in strategically nostalgic fashion to refigure the high summer of Fordism in a way which usefully disrupts the current moral geography of Broadland as a nature region. In its 1965 report, the Nature Conservancy estimated a threefold increase in all craft in the region from 1947 to 1964, with the number of holiday visitors increasing from 100,000 in 1938 to 200,000 in 1955 and 240,000 in 1961. In 1964, 1717 motor cruisers were available for hire (Nature Conservancy, 1965). While boat hire has never been cheap, the effect of this post-war expansion was to move the Broadland holiday down the social scale to the lower middle and upper working class. Until a few years ago, one could find in the more obscure shops of the region tinted postcards from the 1960s and 1970s, communicating this holiday experience with garish cruising images of deck-swabbing, children in the sun, outrageous sunsets, speedboats, lines of cruisers parked for the night, etc.[3] Today's postcards tend more towards birds, mills and reed-cutters, and reprints of late 19th- and early 20th-century photographs.

If this earlier holiday experience is not registered today this is partly because it left little local trace. If travellers write travel books, trippers send postcards and depart, and no formal organizations articulate a collective visitor voice. While boat-owners and hire companies have formal representation, the assumption is that the holidaymaker neither wants nor needs to sustain connection to a region after departure. Postcard and photographic memories are enough until the next time, if there is a next time. Interestingly this temporariness of the holiday connection was challenged in the early 1950s by John May, who attempted to organize the holidaymaker on the basis of loyalty to the region. May's 1952 *Norfolk Broads Holiday Book and Pocket Pilot* asked 'Whose Broads Are They?', and called for an organized visitors' body to challenge closure by private owners: 'What is wanted is a Norfolk Broads Association, with its roots among Broads holiday-makers all over Britain' (May, 1952: 186). Suggesting that visitors might like to return on holiday working parties to clear the waterways of weed, and noting that this might also be a 'movement to raise standards of yachtsmanship and river courtesy' (1952: 187), May argued for a national independent holiday-making voice in regional affairs: 'The Norfolk Broads holiday-maker could become a great power for the good of the Broads if he were organised' (1952: 186).

May's attempt to organize the holidaymaker appears to have come to nothing, the idea of a formal voice perhaps cutting against the grain of a holidaymaking carefree culture of landscape. One has instead to turn to less formal sources to trace elements of a non-improving holiday culture, notably to accounts which celebrate or mock the Broads experience through humour; accounts which presume the regional experience to be inherently a laugh.

Vulgar Fun

It is appropriate to begin here with an ephemeral souvenir of a type now all but extinct in Broadland (Figure 2). Souvenir tea-towels today tend to show scenes of windmills, reeds, sails and birds; have become, to use Lofgren's (1985: 90) distinction in his analysis of postcards, 'scenic' rather than 'comic'. In 1993 I found this rather different souvenir image in a shop in Potter Heigham which has now been demolished. 'We Survived the Norfolk Broads' shows fun with basic facilities in a nature landscape, all cast in garish reds and yellows in the tradition of the seaside postcards with their 'unbearable colours' discussed by George Orwell in his 1942 essay on 'The Art of Donald McGill' (Orwell, 1961). Here is a topographically as well as socially low humour, of people bumping through the scene without a glance at the birds and sky. Eyes down, look in. Even the kite and butterfly are on downward trajectories. The windmill shines over spilt ice cream, children and dogs overboard. A camper dozes in the background, while others in pursuit of nature are caught up in the melee; an angler hooks a woman, a butterfly collector heads for another woman, a hiker is so thirsting for beer that he has lost a boot and is losing more from his torn sack. Those here for a nature region are rendered either cross or absurd, caught up in or oblivious to the real life of the place. Others

Figure 2 'We survived the Norfolk Broads' (souvenir teatowel purchased in Potter Heigham, 1993)

toast the fun with foamy beer. The knitting goes flying as a boat careers with 'Gay Abandon'. And imagine the noise! Are we to conclude that such incidents are to be expected and enjoyed when the kinds of people who purchase such tea-towels are given charge of large waterborne machines? More likely, the towel is to trigger memories of the occasional bit of bad driving in a pleasant environment of pubs and windmills, and/or to operate in terms of what Orwell, in the context of post-cards aimed at the upper working and lower middle class, diagnosed as 'inter-working class snobbery' (Orwell, 1961: 170). Jokes make sense within a moral code having is own respectability, and play on a life less proper which may have been bordered on but not fully embraced through the licence of a week's holiday. The Broads act as licensed premises; there are rules of conduct here too.

This tea-towel is as much part of a culture of landscape as a bird hide. The question is how these different objects become part of competing moral geographies of a nature region – whether nature is a site for self-restraint, self-expression, self-abandonment, etc. – and of how these different modes of being become memorialized in different ways, if at all. If it is difficult now to purchase such memorial objects as 'We Survived the Norfolk Broads', there are occasional glimpses of this form of non-improving Broadland heritage in the contemporary landscape. The saucy postcard ethos of the tea-towel connects with an earlier celebrity visitor to the region, variety and film star George Formby, who, from 1947, in between appearances in Yarmouth's summer season, would cruise every year with his wife on the Broads (Randall and Seaton, 1974). Formby owned a riverside home near Wroxham, and in 1991 Norfolk County Council sought to demolish 'Heronby' as part of a bypass scheme. Protesting George Formby Society members 'took along their ukeleles and sang Formby classics on the riverside behind the boarded-up luxury home' (*Eastern Evening News*, 10 June 1991). Another kind of heritage intrudes into the Broadland scene.

The saucy sensibility evinced by Formby and the tea-towel can be traced in Broadland throughout the past 100 years, the tea-towel perhaps being its fag end. If the tea-towel is a visual product, the way of being it depicts and the manner of its use resist capture by the idea of the 'tourist gaze' (Urry, 1990). The purchased object, whether used to wipe dishes or pinned upon the wall, works through a triggering of more than visual memories. Sights may be brought to mind, but also smells and sounds and textures, and it would lose purchase on this object to consider individual senses in isolation. 'We Survived the Norfolk Broads' concerns the holiday as corporeal incident, as indeed does Ransome's *Coot Club*. The opposition is less between gazing children and vulgarly corporeal Hullabaloos than between two modes of corporeality in landscape, distinguished by the form and degree of sensual coordination rather than the triumph of one sense over another, with each denoting the production of environmental knowledges through what Crang terms their different 'occurrent landscapes' (Crang, 1999: 246).[4] And if these corporealities differ in ways analogous to Urry's (1990) distinction between 'romantic' and 'mass' visual tourism, the latter here denotes a sensibility which its critics could detect across classes. The moral geographies of Broadland concern a sensibility claiming the virtues of the respectable middle class, reaching through an educational ethos to lower social groups and allying itself with responsible upper-class landownership, setting itself against an unholy Hullabaloo alliance of the working class and upper-middle and upper class. Vulgarity is not only ascribed to the socially low. That at least would be the view from within a sensibility which finds its latest institutional manifestation in the Broads Authority. From other perspectives these social categories might fracture or coalesce in different ways, in each case refracting different cultures of Englishness through the region; contemplative romanticism, responsible ruralism, seaside vulgarity.

Vulgarity, then, has acted both as a defining other for an improving culture of nature, and as something to be celebrated in its own right as part of what Lofgren terms a 'subtle cultural warfare' (1985: 98) stressing bodily impropriety. If vulgarity is here a 'double-sided weapon' (Lofgren, 1985: 103), a term deployed both to hail and put down a particular culture of landscape, at times the two sides have met, notably in the guides to Broads conduct issued by the leading boat hire agency, Blakes. In 1922 Blakes issued *Broad Smiles: Or How Not to Do a Norfolk Broads Holiday* (Blakes, 1922), a book of cartoons caricaturing every social leisure group; 'motor hogs', lecherous men watching things other than birds through binoculars, jazzing youths, incompetent yachtsmen. *Broad Smiles* complemented Blakes' yearly handbook *How to Do the Norfolk Broads and What to Take*, but the relationship between the two publications was not a straightforward one of propriety and impropriety; rather *Broad Smiles* could be read by the Broads novice as a jocular but tempting document – 'If you are thinking of spending your next holiday in glorious Broadland our artist has exaggerated a few natural happenings for your amusement' – and by the Broads enthusiast with a smiling self-recognition, both in terms of having seen such things go on and having been party to them: 'If you experienced the delights of a Norfolk Broads Holiday this little book will not need an introduction' (Blakes, 1922: title page). Such guides to Broadland as a site of japes are a staple of the regional literature through the century, acting to some degree as a thoroughly unsubversive cultural safety valve, 'a sort of saturnalia, a harmless rebellion against virtue' (Orwell, 1961: 178), but also signalling that such goings on are a fixture of the scene, whose celebration through humour can undercut more serious moral geographies of Broadland as improving space.

In 1948 Blakes published *Let's Be Broad-Minded!*, a 'Bunk-Side Book of Brighter Yachting', written by Dennis Rooke and illustrated by Alan d'Egville (Rooke, 1948); the two were known for producing comic illustrated books. The exclamation mark in the title is indicative of a hybrid *Punch*/end-of-the-pier sensibility wandering through a nature region, providing a mock regional geography equivalent to Sellar and Yeatman's mock British history *1066 and All That* (Sellar and Yeatman, 1930). Had the future *Carry On* ... film series included *Carry on Sailing*, or (more likely) *Carry on Broading*, Rooke's text and d'Egville's cartoons would have offered a proto-screenplay. The book purports to recount a recent sailing holiday with two women ('Beautiful Ladies') on the boat *Beautiful Lady*, 'portrayed so faithfully (well sometimes, anyway)'. Having described the Broads as a safe sailing environment, Rooke rhetorically asks whether anything happens there at all: 'As a matter of fact there is no need to fear. Amusement *is* to be had. Incidents *do* take place. It is only a matter of time ...' (Rooke, 1948: 9).

Rooke suggests you may fall in while quanting, inadvertently lasso a man mowing his lawn, or hit your head going under Potter Heigham Bridge. A ship's log gives an embroidered story of their week on board, including one party member's antiquarian passion for walking to 'q. t. f.-t.' churches (quaint thatched, flint-towered) (Rooke, 1948: 93). Those who seek to gaze upon objects scripted as authentically historic, those who connect to place only through visualization, are satirized, in a strange anticipation of Roland Barthes' 1957 satire on the *Guide Bleu* vision in *Mythologies* (Barthes, 1973).[5] Various other Broads types are paraded; the 'running commentator' who has been before and insists on saying what is round the next corner, the 'nautical-vocabulary critic' of your sailing manner and knotting ability, the 'fashion-parader', the 'long-nailed glamour-girl', etc. Illustrations show various outlandish poses and gestures, with occasional hints of nudity and 'broad-minded!' goings-on. The assumption throughout is that cruising revolves around the consumption of alcohol. This is perhaps what the Broads Authority have in mind when they write that: 'In the past, some holiday companies have marketed Broads holidays with scant regard for the unique character and fragile nature of the area, encouraging visitors to come to the Broads for the wrong reasons' (Broads Authority, 1989).

Theatreland

Rooke and d'Egville present the Broads as a theatrical space, a right performance, and indeed the metaphor of theatre recurs in the more saucy, non-improving versions of Broadland, allowing commentators to retain the authority of a surveyor of the scene by acting as authorial compère, and integrating life and environment in a way rather different to that of Ransome or the Broads Authority. Landscape and nature play just as crucial a role, but to very different effect.

It is important here to distinguish this specific metaphor of theatre from a general language of performance. If Nash (2000) suggests in the context of work on dance that we should respect the specificity of that practice by considering it in terms of choreography rather than gathering it under a general rubric of performativity, so here we should be attentive to the specific modes of theatricality working in the region, which generate particular configurations of regional subject and object. A sense of Broadland as theatre is most fully developed in the writing of Drew Miller, a retired American journalist and banker, keen sailor and the author of several plays and short stories, who, until his death in 1936, published yearly guides to *What to do on the Norfolk Broads* from his converted mill home on the banks of the Bure near Horning (*East Anglian Magazine*, 1936). Miller's last book, *Seen from a Windmill: A Norfolk Broads Revue* (Miller, 1935), opens with 'The Curtain Rises', introducing the author and various friends who feature in the

Figure 3 'The Show Begins' (opening image from Miller, 1935)

subsequent tales. Theatricality allows Miller to cross between the worm's-eye view of the characters placed within the theatre of regional space, and the overview of an authorial director or compère. The specific genre of revue allows him to move a variety or music hall sensibility more conventionally at home in Great Yarmouth into Broadland. An opening image, 'The Show Begins' (Figure 3), shows Miller's mill as a lighthouse illuminating the Broads stage, the author's pen bringing out its many stories: 'The famous thousand and one tales of Scheherazade are insignificant in volume compared with the vast number of stories floating about in Broads country' (Miller, 1935: 13). Styling the Broads as a theatrical narrative space is in part a matter of recounting ghost stories and yarns spun by old residents, but it also concerns the world of the visitor: 'People, gliding by in boats, may seem prosaic and uninteresting, but hidden behind their phlegmatic stolidness can often be found the most entrancing romances.' Miller notes of the region that: 'In the beginning it was its unique beauty that appealed, but, as I came to know them, it was its people and their stories that proved the more interesting' (Miller, 1935: 14).[6]

Miller's friends are wary of his writing, fearing that it may attract more visitors and accentuate 'the jazzing of the Broads' (Miller, 1935: 13), yet Miller shows them also relishing raunchy behaviour, with Old Debbage the wherryman licking his lips when recalling the flagrant antics of a young cruising party. In Miller's stories such visitors appear as an integral part of a rich narrative tapestry, with one of the key threads of that tapestry being saucy tales of older men and younger women. 'The Adventure of the Amorous Colonel' concerns an old gent taking a young blonde for a cruise aboard the *Love Nest*, only to be tracked down by his wife. 'The Adventure of the River Censor' concerns a self-important man, 'Old Bean', who disregards the rules of the river when visiting in his own speeding boat, and mixes hypocrisy and prurience in his quest 'to maintain the Broads on a high moral plane' (Miller, 1935: 112). This 'self-appointed censor of the Bure' shouts at speeding craft and attacks certain modes of behaviour:

> He wrote letters to the newspapers denouncing the modern bathing suit, and, in order to be fully versed as to what was or was not being worn, he would use high-powered binoculars. These enabled him to make many interesting discoveries, and provided him with material for some luscious articles on the subject of 'Nudism on the Broads'. These were eagerly read by many, who got great enjoyment from the shocks they received.

Old Bean spends his evenings flashing a searchlight into secluded waterways, 'causing no little confusion to the amorous' (Miller, 1935: 112).

Miller provides an allegory of others seeking to regulate the Broads' moral geography via definitions of the right and wrong ways to love the region. Miller is no libertarian, and can lament speeding and jazz along with the next man, but his Broads revue offers a rather different way of telling the region, and hints at further cultural complexities of conduct in nature, not least regarding the sexual body.

Overbalance: Freedom and its Limits

> Young men who lounge in a nude state on boats while ladies are passing (and I have known Norfolk youths to do this) may be saluted with dust shot, or the end of a quant. (G.C. Davies, quoted in Arrow, 1951: 22)

The reader should not imagine that this article has been written with a view to policy recommendation; to suggest that Broadland should be noisier than it is, or that the Broads Authority should institute compulsory transistor or ukelele playing to liven the place up. Rather, the version of Broadland as a space of non-improving fun has been raised to illuminate the moral geographies and histories that run through the currently dominant sense of Broadland as nature region, and to indicate other elements in the region's cultural grain.

Questions of morality and freedom are central to any leisure landscape. The issue of planning for leisure, and of designating spaces such as national parks as sites for valued forms of leisure, entails judgements over how 'free' time should play over space, and how people might best use their play for their own good and for the good of the environment which they are in. Such judgements and tensions are inherent in and formative of landscapes of leisure and pleasure (Clark et al., 1994; Matless, 1997b). In Broadland, the culture of landscape turns on attempts to balance freedom and regulation through the elevation of a quiet and educational encounter with the natural world. The liberal balance of freedom and regulation through education has, however, been tipped each way, either by those seeking stronger controls on visitor activity, or by those resisting control in the name of liberty in landscape. The most strident representative of the former stance in 20th-century Broadland was James Wentworth Day, author and sportsman, who shot and pronounced over the region from the late 1930s to the 1970s (Matless, 2000). Day regarded the tripper as a form of ecological and social scum, destroying the place through their exercise of bad freedom, including sexual licence. In his 1967 popular topographic book, *Portrait of the Broads*, Day remembered a Broadland not 'exploited, capitalized, vulgarized, transistorized' (Day, 1967: 16): 'A pleasanter picture I suggest than that of a boatyard yacht-basin littered with ice-cream cartons, candy floss sticks, empty potato crisp bags, cigarette ends and the thin grey scum of contraceptives on the water' (1967: 36).[7] A self-styled maverick reactionary, Day diagnoses the permissive society as an enemy of local nature, with sexual licence as part of a bad modern freedom to consume.

If Day sought to proscribe freedom via a restrictive morality, others have challenged the educational balance of regulated freedom from the other side. The quotation from Davies on nude display at the head of this section is taken from an anthology on sailing by John Arrow, secretary of the Broadland Protection Society in the late 1940s,[8] and author of a proposal for 'The Broads as a National Park' in the *Architectural Review* in 1949. Arrow's proposal began:

> The more the English become an urban race the more necessary it is to ensure that a counterbalance to their urban environment is available in the shape of tracts of wild countryside where they can breathe deeply (both literally and metaphorically) and where they can enjoy a life in which their own initiative has full scope in contrast to the synthetically organized life of the town. (Arrow, 1949: 87)

Arrow went on to develop arguments on sailing, self-reliance and self-fulfilment; two decades earlier he had given such pursuits of freedom a rather different inflection in his novel, *Young Man's Testament* (Arrow, 1932). Here Arrow took his message on the synthetic nature of urban life and the freedoms of nature from D.H. Lawrence, whom he had recently defended against accusations of pornography (Arrow, 1930), and presented the Broads as a site for sexual self-realization in a way which overbalanced any liberal blend of freedom and regulation. Arrow describes three couples taking a cruiser from Essex up the coast to find the Broads, and each other. The Broads landscape becomes a key character in a moral geographic tale which leaves Davies's injunction to quant the naked far behind. Here the Broads are a space of moral freedom; of a freedom from conventional morality, yet a freedom which is of a higher morality. Boat-owner Jocelyn, personification of a desire for order through his organization and control of the boat, loses his sensualist girlfriend Cecily to Mark, who has been held back by his repressed girlfriend Gertrude, who in turn falls for Jocelyn. Cecily's sister Thelma, and her boyfriend Jim, share a cabin as lovers at the start of the book, but become increasingly distant from one another. The micro-geographies of the boat and the open spaces of the region are played off, as cabins are swapped and the changed pairings find each other through differential encounters in the landscape: 'by imposing its restriction of space, the boat had accelerated the emotional changes that had lain dormant' (Arrow, 1932: 177; cf. Crouch, 1999 on the caravan).

On the return journey to Essex Thelma responds to Jim's unwanted advances by pushing him overboard at sea, where he drowns; the group decide that such an action is understandable given provocation, and report the incident as an accident. Mark and Cecily are the heroic figures of the book, coming together through a sensual delight in Broadland as a nature space, and realizing both Broadland and themselves through open-air sex at Potter Heigham. In Lawrentian style Cecily realizes herself as a woman through submission to a dominant male, and dances naked for Mark in a field, only to be disturbed by walkers: '"Do you think they saw?" "No, no, they were deep in conversation about bacon, or something"' (Arrow, 1932: 170). By contrast Gertrude, representative of conventional values and against sex before marriage, is 'an absolutely indefatigable sightseer', and likes postcards: 'To-day she had bought some post cards. She rather liked them. They were photographs of the broads, produced with a refined and devastating sterility' (Arrow, 1932: 141). The tourist gaze is dramatized as a repressed engagement with place, a means of asserting mind over body, much as in *Broad Smiles* the q.t. f.-t. church fan could not enjoy the region for sightseeing. Gertrude and Jocelyn are:

> . . . the examples of over-civilization set down in an environment almost utterly pagan. . . . Space existed all around them, they were free; but they shut themselves within the limitations of their minds and disowned their bodies in obedience to a stifling convention. (Arrow, 1932: 188)

Broadland as a nature space of freedom demands a throwing off of conventional conduct, and an embrace of the body.

Concluding Comments

There is much more that could be drawn from Arrow's novel; here *Young Man's Testament* indicates how a balance of freedom and regulation could tip in the direction of the former in the name of a moral geography far from either the comic tea-towel or the Coot Club. In playing Lawrentian themes of sexuality, masculinity and femininity through landscape, Arrow also demonstrates the presence of a theoretical literacy regarding body and society within the region over the past century. If the moral geography of Broadland is a story of action and noise over 100 years, it is also a story of theoretically and politically reflexive people whose arguments at times anticipate more recent concerns for the body and its spaces. This may not be surprising with the Lawrentian Arrow, but I would suggest it holds for all of the figures encountered in this article: Miller with his theatrical sauce, Ransome with his observant children, Wentworth Day with his maverick critiques of commerce as a barrier to intimate place-knowledge, the anonymous artist of the tea-towel. All are the authors of imaginative and reflexive action, much like any visitor or resident of the region. Rather than take a particular theoretical perspective on body and society as a means to judge the discourses and seeming prejudices of its subjects as historically contextual, classifiable and containable, this article has sought to present the region as something generated through theorizations, practices and arguments of body and society on the part of the people within it, whether resident or visitor. Such a genealogy of place may unsettle assumptions concerning both the provenance of theory and the make-up of a region claimed for Nature.

The dominant culture of nature in contemporary Broadland seeks to balance mind and body with neither one dominating the other, and to balance freedom and regulation through an observant care for the region, itself under the care of the Broads Authority. A cultural distance is kept from another popular culture of Broadland as a space of non-improving fun; the waterland drives back the seaside. If the landscape of contemporary Broadland reflects the relative demise of the latter, two images from July 1995 of its continuing minor presence can serve to end these reflections on landscape and cultural value. Both are drawn from a showpiece conservationist site, both are minor acts registering different desires. One concerns an act of popular mourning, the other an act of popular trashing.

In the upstairs gallery of the Norfolk Wildlife Trust's thatched Conservation Centre at Ranworth, children had been invited to draw crayon pictures of what they had seen; images of herons, ducks and beetles covered the wall, reflecting their experience of a day out. In the middle of this nature display one drawing showed the sign for the popular boy band, Take That, with a crayoned plea to their recently departed member Robbie Williams: 'Come back Robbie'. Nature could not put pop grief out of one child's mind.

In the downstairs gallery, a 'Litter Can Kill' exhibit included a deflated 'Mr Blobby', the pink and yellow-spotted sidekick of TV celebrity Noel Edmonds, and at the time something of a cult among children and adults alike. The binning of Blobby was somehow symbolic of all the Broads Authority had been trying to achieve, and the Coot Club would have been proud. Blobby lay, less than potent, among cans and fishing line and hooks and mess; a plastic thing out of place, something that should never have come to the Broads. Children and adults alike should take such rubbish home.

Notes

Part of the research for this article was funded by a University of Nottingham grant (94NLRG454). Thanks also to anonymous referees for insightful comments.

1. Rose is here contrasting spatialization with 'the apparent linearity, unidirectionality and irreversibility of time' (1996: 144). I would extend Rose's sense of the complexity of space to the temporal and historical. I take the metaphor of cultural 'grain' from William Carlos Williams' *In the American Grain* (1989), a retelling of history which draws out different and often conflicting elements of 'the American' by re-inhabiting various historical narratives. In a way, the current article aims to do a similar thing for Broadland, if in a different narrative form.

2. Nash's essay is primarily concerned with Nigel Thrift's recent work on dance, non-representational theory and the ecology of place, though Nash perhaps misses Thrift's (1999) attention to a Wittgensteinian notion of language as practice. Thrift expresses a reasonable antipathy to certain forms of cultural geography which have modelled landscape as text and/or focused exclusively on issues of representation, but the force of this criticism perhaps leads him to sustain distinctions of representation (as represented by certain cultural geographers, notably James Duncan) and practice which other elements of his work cut against. It should be clear that this article is not seeking to model the Norfolk Broads as 'text'.

3. Art photography has begun to revisit such images as part of the rediscovery of colour photography as a mechanism for catching the quotidian, or for the staging of ironic tableaux of everyday life. See, for example, the reclamation of the work of the late John Hinde, founder of the Hinde postcard company (Hinde, 1993). Martin Parr has supplemented his colour investigations of British class and culture with a collection of *Boring Postcards* (Parr, 1999) commemorating such glories of the British modern as motorway service stations, shopping precincts and holiday camps. Such modern images have yet to be reissued in Broadland. On postcard cultures see Lofgren (1985).

4. Crang's 'occurrent landscape' refers to the sense of events waiting to happen within the semiotic economy of tourism, and takes tourism as a reflexive and imaginative practice, a self-knowing operation (see also Crang, 1997).

5. Barthes argued of the mythologist that, 'condemned to live in a theoretical sociality. . . . His connection with the world is of the order of sarcasm' (Barthes, 1973: 157). Rooke may not occupy a theoretical sociality, but in his pop sociology, too, sarcasm functions as the grain of truth. I am not suggesting here that Rooke is somehow an equivalent of Barthes, but it would be instructive to place Barthes in relation to the wider field of social geographical satire, as well as in the context of cultural and political criticism in the period (see Rylance, 1994: 54–65 for a comparison with British figures such as Hoggart and Williams).

6. Miller's theatrical openness to Broadland stories does not prevent him deploying crude ethnic stereotypes in describing visitors, as in his story of 'The Adventures of the Harassed Hebrews' (Miller, 1935: 151–8).

7. Martin George (1992: 122) notes that the discharge of faeces and contraceptives directly into rivers and broads from boat toilets caused a 'major aesthetic problem' at busy sites in the 1960s, before regulations were introduced prohibiting the direct discharge of boat sewage in 1971.

8. The Broadland Protection Society was founded by boat-builder and hire operator Herbert Woods, and was praised by Rooke (1948: 104) and May (1952: 183), both of whom highlighted its efforts to open up private broads to navigation, including through direct action. May argued, however, that it could never be more than the voice of the boat-owners, and needed to be complemented by an organization directly representing the boat-user.

References

Arrow, John (1930) *J.C. Squire v D.H. Lawrence: A Reply to Mr.Squire's Article in 'The Observer' of March 9th, 1930*. London: Blue Moon Booklets.

Arrow, John (1932) *Young Man's Testament*. London: Putnam.

Arrow, John (1949) 'The Broads as a National Park', *Architectural Review* 106 (632): 86–100.

Arrow, J. (ed.) (1951) *The Pleasures of Sailing: an Anthology*. London: Art and Technics.

Bachelard, Gaston (1983) *Water and Dreams*. Dallas, TX: Dallas Institute of Humanities and Culture.

Barthes, Roland (1973) *Mythologies*. London: Paladin.

Blakes Ltd (1922) *Broad Smiles. Or How not to Do a Norfolk Broads Holiday*. London: Blakes Ltd.

Broads Authority (1989) *The Broads . . . Last Enchanted Land*. Norwich: Broads Authority (leaflet).

Broads Authority (1993) *No Easy Answers: Draft Broads Plan 1993*. Norwich: Broads Authority.

Brogan, Hugh (1984) *The Life of Arthur Ransome*. London: Jonathan Cape.

Cameron, Laura (1997) *Openings: A Meditation on History, Method, and Sumas Lake*. McGill: Queen's University Press.

Cameron, Laura (1999) 'Histories of Disturbance', *Radical History Review* 74: 4–24.

Clark, M. Aitken (1989) 'Britain's Newest and Very Special National Park', in R. Denyer *Still Waters*. Norwich: Still Waters Press.

Clark, G., J. Darrall, R. Grove-White, P. Macnaghten and J. Urry (1994) *Leisure Landscapes*. London: CPRE.

Corbin, Alain (1994) *The Lure of the Sea*. Cambridge: Polity.

Crang, Mike (1996) 'Envisioning Urban Histories: Bristol as Palimpsest, Postcards and Snapshots', *Environment and Planning A* 28: 429–52.

Crang, Mike (1997) 'Picturing Practices: Research through the Tourist Gaze', *Progress in Human Geography* 21: 359–74.

Crang, Mike (1999) 'Knowing, Tourism and Practices of Vision', pp. 238–56 in D. Crouch (ed.) *Leisure/Tourism Geographies*. London: Routledge.

Cresswell, Tim (1996) *In Place/Out of Place*. London: University of Minnesota Press.

Crouch, David (1999) 'The Intimacy and Expansion of Space', pp. 257–76 in D. Crouch (ed.) *Leisure/Tourism Geographies*. London: Routledge.

Davies, G. Christopher (1923) *Handbook to the Rivers and Broads of Norfolk and Suffolk*. London: Jarrold. (Orig. pub. 1882.)

Davies, G. Christopher (1932) *The Swan and Her Crew*. London: Methuen. (Orig. pub. Frederick Warne, 1876.)

Day, James Wentworth (1967) *Portrait of the Broads*. London: Robert Hale.

Denyer, Richard (1989) *Still Waters*. Norwich: Still Waters Press.

Driver, Felix (1988) 'Moral Geographies: Social Science and the Urban Environment in Mid-Nineteenth-Century England', *Transactions of the Institute of British Geographers* 13(3): 275–87.

East Anglian Magazine (1936) 'East Anglian Authors – No.2 – A.G.Miller', *East Anglian Magazine* 1(9) June: 514.

Ewans, Martin (1992) *The Battle for the Broads*. Lavenham: Terence Dalton.

Foucault, Michel (1986) *The Use of Pleasure*. Harmondsworth: Penguin.

George, Martin (1992) *The Land Use, Ecology and Conservation of Broadland*. Chichester: Packard.

Giblett, Rod (1996) *Postmodern Wetlands: Culture, History, Ecology*. Edinburgh: Edinburgh University Press.

Goubert, Jean-Pierre (1989) *The Conquest of Water*. Cambridge: Polity.

The Graphic (1893) 'A Week in a Wherry', *The Graphic* 28 October: 544–5.

Hinde, John (1993) *Hindesight: Photographs and Postcards by John Hinde Ltd. 1935–1971*. Dublin: Museum of Modern Art.

Illich, Ivan (1986) *H2O and the Waters of Forgetfulness*. Dallas, TX: Dallas Institute of Humanities and Culture.
Knights, Sarah (1989) 'Representation and Reality: A View of the Broads', in R. Denyer *Still Waters*. Norwich: Still Waters Press.
Laurier, Eric (1998) 'Replication and Restoration: Ways of Making Maritime Heritage', *Journal of Material Culture* 3: 21–50.
Laurier, Eric (1999) 'That Sinking Feeling: Elitism, Working Leisure and Yachting', pp. 195–213 in D. Crouch (ed.) *Leisure/Tourism Geographies*. London: Routledge.
Lofgren, Orvar (1985) 'Wish You Were Here! Holiday Images and Picture Postcards', *Ethnologia Scandinavica*: 90–107.
Mackenzie, John (1988) *The Empire of Nature: Hunting, Conservation and British Imperialism*. Manchester: Manchester University Press.
Macnaghten, Phil and John Urry (1998) *Contested Natures*. London: Sage.
McWilliam, Neil and Veronica Sekules (eds) (1986) *Life and Landscape: P.H. Emerson, Art and Photography in East Anglia 1885–1900*. Norwich: Sainsbury Centre for Visual Arts.
Matless, David (1992) 'A Modern Stream: Water, Landscape, Modernism and Geography', *Environment and Planning D: Society and Space* 10: 569–88.
Matless, David (1994) 'Moral Geography in Broadland', *Ecumene* 1(2): 127–56.
Matless, David (1997a) 'The Geographical Self, the Nature of the Social and Geoaesthetics', *Progress in Human Geography* 21: 393–405.
Matless, David (1997b) 'Moral Geographies of English Landscape', *Landscape Research* 22(2): 141–55.
Matless, David (1998) *Landscape and Englishness*. London: Reaktion Books.
Matless, David (2000) 'Versions of Animal-Human: Broadland, c.1945–1970', in C. Philo and C. Wilbert (eds) *Animal Spaces*. London: Routledge.
May, John (1952) *The Norfolk Broads Holiday Book and Pocket Pilot*. London: Hulton Press.
Miller, Drew (1935) *Seen from a Windmill: A Norfolk Broads Revue*. London: Heath Cranton.
Nash, Catherine (2000) 'Performativity in Practice', *Progress in Human Geography* 24.
Nature Conservancy (1965) *Report on Broadland*. London: Nature Conservancy.
Nicholson-Lord, David (1987) *The Greening of the Cities*. London: Routledge.
Orwell, George (1961) 'The Art of Donald McGill', pp. 167–78 in *Collected Essays*. London: Secker and Warburg.
Parr, Martin (1999) *Boring Postcards*. London: Phaidon.
Randall, Alan and Ray Seaton (1974) *George Formby*. London: W.H. Allen.
Ransome, Arthur (1940) *The Big Six*. London: Jonathan Cape.
Ransome, Arthur (1969) *Coot Club*. Harmondsworth: Puffin. (Orig. pub. Jonathan Cape, 1934.)
Rooke, Dennis (1948) *Let's Be Broad-Minded! The Bunk-Side Book of Brighter Yachting*. London: Blakes Ltd.
Rose, Nikolas (1996) 'Identity, Genealogy, History', pp. 128–50 in S. Hall and P. Du Gay (eds) *Questions of Cultural Identity*. London: Sage.
Rosenthal, Michael (1986) *The Character Factory*. London: Collins.
Rylance, Rick (1994) *Roland Barthes*. Hemel Hempstead: Harvester Wheatsheaf.
Samuel, Raphael (1994) *Theatres of Memory*. London: Verso.
Sellar, W.C. and R.J. Yeatman (1930) *1066 and All That*. London: Methuen.
Springhall, John (1977) *Youth, Empire and Society*. London: Croom Helm.
Stallybrass, Peter and Allon White (1986) *The Politics and Poetics of Transgression*. London: Methuen.
Taylor, John (1995) *A Dream of England: Landscape, Photography and the Tourist's Imagination*. Manchester: Manchester University Press.
Taylor, Harvey (1997) *A Claim on the Countryside: A History of the British Outdoor Movement*. Keele: Keele University Press.

Thacker, Andrew (1993) 'Foucault's Aesthetics of Existence', *Radical Philosophy* 63: 13–21.
Thrift, Nigel (1999) 'Steps to an Ecology of Place', pp. 295–322 in D. Massey, J. Allen and P. Sarre (eds) *Human Geography Today*. Cambridge: Polity.
Urry, John (1990) *The Tourist Gaze*. London: Sage.
Urry, John (1995) *Consuming Places*. London: Routledge.
Wardale, Roger (1988) *Arthur Ransome's East Anglia*. North Walsham: Poppyland Publishing.
Williams, William Carlos (1989) *In the American Grain*. Harmondsworth: Penguin. (Orig. pub. 1925.)
Williamson, Tom (1997) *The Norfolk Broads: A Landscape History*. Manchester: Manchester University Press.
Wilson, Alexander (1992) *The Culture of Nature*. Oxford: Blackwell.
Wolff, Michael and Celina Fox (1973) 'Pictures from the Magazines', pp. 559–82 in H.J. Dyos and M. Wolff (eds) *The Victorian City: Images and Realities, vol. 2*. London: Routledge.
Wright, Patrick (1999) *The River: The Thames in Our Time*. London: BBC Books.

David Matless is Reader in Cultural Geography at the University of Nottingham. His publications include *Landscape and Englishness* (Reaktion Books, 1998) and *The Place of Music* (London and New York: Guilford, 1998, with A. Leyshon and G. Revill).

Bodies in the Woods

PHIL MACNAGHTEN AND JOHN URRY

Bodies and Trees

Trees figure strongly in this *Body & Society* collection concerned with how bodies are in, and of, 'nature'. Many commentators more generally presume that there is something 'natural' about being in the presence of trees, and that certain kinds of tree or arrangements of trees bring one's body into a closer relationship to the natural world. Many writers presume that humans exhibit a general affinity with trees, and that trees culturally signify a universal 'natural' message.

But although there is something about trees and bodies that does always seem to effect a passionate liaison, the nature of this liaison varies from society to society. Powerfully different social myths have been woven into the roots and branches of various cultures (see Schama, 1995). Thus in America, the discovery of the Big Trees of California was seen as an American godsend, the revelation of the uniqueness of America and of the particularly chosen character of the American people. Preachers, painters, photographers and writers described and lyricized its stupendous sequoias and their apparently God-given character. In Germany forests have long been viewed as representing the spirit of militarism; an embodied memory that the modernizing Nazis deployed and developed. Goering turned huge tracts of forest into magnificent hunting grounds; forest themes invaded most aspects of Nazi art and politics; Hitler and his fellow Nazis were regularly photographed in woodland settings; and extensive programmes of woodland protection were introduced. Large areas of Poland were subjected to a total landscape plan in which villages were depopulated and the land turned into hunting forests, although in Poland forests had long stood for the enduring struggle for national freedom against various outside invaders. In France forests

Body & Society © 2000 SAGE Publications (London, Thousand Oaks and New Delhi),
Vol. 6(3–4): 166–182
[1357–034X(200009/12)6:3–4;166–182;015476]

have represented the passion for order and rational planning by the state, while in England the forest has stood for the idea that liberty against a despot could be attained by those living under the greenwood tree (as in the Robin Hood myth).

New meanings and uses of woods and forests have often accompanied patterns of colonial conquest. When the British colonized India they first took over the forests and displaced local rights, needs and indigenous forms of knowledge (Shiva, 1989). The religions and cultures of South Asia had all been rooted in the forests, and they were seen as places of dwelling where 'humans and nature' were indissolubly linked. The complex and extensive patterns of indigenous forest management, where trees conserved soil and water as primary sources of life, were soon transformed under colonial government as forests were turned into plantations of 'timber mines' based upon the newly emergent science of forestry.

Tree symbolism has been the subject of considerable anthropological inquiry. Across numerous cultures and societies trees are used as symbols of transgenerational continuity; the vitality and self-generation of trees symbolic of 'life' itself (see Rival, 1998). Rituals marking the life process often make extensive use of trees. In Bali the coconut is a central ingredient of birth, marriage and death (Giambelli, 1998); in Romanian folklore fruit trees are associated with death and marriage rituals (Rival, 1998); in south India the coconut tree is planted on the grave representing the life force of the dead (Uchiyamada, 1998); while in Papua New Guinea the red pandanus tree is used in male initiation ceremonies (Bonnemere, 1998). The endurance of trees can also express patterns of social continuity and stability. Japanese culture is often referred to as a tree culture or '*ki no bunka*' (Knight, 1998). Trees and wood are very much part of everyday life, especially in the traditional Japanese *ei* or stem family, where the focus of concern is to tie the fortunes of the *ei* to long-term patterns of care and nurturance; in other words to the transgenerational rhythms of trees and woods rather than that of individual lifetimes. In south India different trees symbolize different forms of continuity: while the banyan tree represents the village community as a whole, the coconut palm symbolizes the life cycle of individuals (Giambelli, 1998).

And in contemporary contestations of nature, trees continue to play a particularly significant role (see Henwood and Pigeon, 1998; Macnaghten and Urry, 1998). Thus ancient trees are often seen as signifiers of the long-established naturally occurring organic communities that had grown up alongside those trees (see Ingold, 1993). The seasonality of trees, with leaves growing and then falling, can be viewed as denoting the natural seasons which modern, urban societies have threatened to destroy. Protests against the cutting down of trees come to signify wider 'unease' with industrial modernity. While trees are identified with humans, and admired for their longevity, their great proportions, potency and self-generating energy, this is contrasted to all that appears false and superficial and instantaneous in consumer-oriented society (Rival, 1998). Trees stand in opposition to manufactured goods, spectacularly those of arms and cars, the symbols of death and decay (Zelter, 1998). Hence the considerable numbers of people who dangerously place their own bodies in the way of earth-moving equipment so as to save the body of the tree, especially from new roads and airports. Many seem to feel particular affinities with trees since, like the upright human body, they appear majestically defenceless against progress, the modern and the scientific. Road protesters, hugging their adopted tree trunks for dear life, appear naturally entwined, trunk to trunk, arms, legs and branches, in a passionate embrace of living bodies (see Macnaghten and Urry, 1998: 70). Those trees that have been subject to acid rain are especially emotive icons of contemporary environmentalism. The twisted, warped, gnarled remains of once-proud trees bear witness to the power of human hubris to generate environments where the moving, sensuous natural human body is, for the foreseeable future, hugely out of place. Certain kinds of trees can be seen as awesomely unnatural in the landscape. As we explore later, the ranks of conifers 'marching up to the top of the hill and marching down again' are one such example in contemporary Britain. Bodies in such forests do not imbibe the energizing morality inspired by a closeness with nature. They need to be disciplined, regulated and unnatural, as dangerous intrusions that, if unregimented, might cause the industrialized 'unnatural' rows of trees to go up in smoke.

However, for all these general claims about trees and humans, what is noteworthy is that there is little or no research into how specific social groups do in fact *engage with and perform their bodies in* different kinds of wooded environment in the West. It seems that people experience being in woods and forests as 'natural' and that walking, hunting, picnicking, running and so on through trees bring the body closer to nature, but these claims have not been established through systematic research. It further seems likely that different social groups will experience quite differently the bodily opportunities *and* bodily constraints that woods and forests provide, but again this can only be established by detailed research.[1]

The next sections of this article report on research that we conducted in Britain to try to begin to resolve some of these issues, especially with regard to the leisure landscapes of woods and forests (this is therefore a continuation of themes outlined in Clark et al., 1994). We researched just how people seek to dwell in relationship to woods and forests, how their relationships are sensuously embodied and how they provide different opportunities for expressing people's apparently 'real nature', outside in the 'open air'. We make some use of the concept of affordances to understand these processes. Gibson argues that in the environment we do not encounter a set of objective 'things' that may or may not be visually perceived, but rather we encounter different surfaces and different objects, relative to the human organism, and these provide affordances (1979: see the Introduction in this volume). Affordances are both objective and subjective, both part of the environment and part of the organism. They stem from the reciprocity between the environment and the organism, deriving from how people are kinaesthetically active within their world (Costall, 1995: 475). Affordances constrain behaviour along certain possibilities, connected to bodily capacities and limits of the human organism (albeit augmented by various objects and technologies: boots, coats, compass, food and so on). Some such affordances are a path that beckons people to walk safely along it, a wood that is a repository of childhood memories, trees that invite young children to climb and so on. Humans in woods and forests are active, vital, corporeal and mobile beings that are afforded various and conflicting possibilities.

We shall show that this research demonstrates first, that there are significant, contested and ambivalent affordances provided by woods and forests in contemporary Britain. Second, there is considerable variation in the embodied experiences of trees between different social groups; there are, we might say, different 'contested natures' in the forest. And, third, those organizations concerned with 'managing' such places deal problematically with the embodied relationship that some groups have with trees, woods and forests. Some disciplined bodies are also, we show, resistant bodies.

Outside and Closer to 'Nature'[2]

This research consists of nine focus groups, held at various locations in Scotland, Wales and England in 1998. These groups reflected a spectrum of social class, ethnic backgrounds and age groups. In the following extensive quotation, male participants are 'M', female participants 'F' and the moderator 'Mod'.

Getting out into the 'open air' in general is very important in most people's daily lives. It offers scope for relaxation, to refresh oneself bodily, and to re-form social relationships. There was much discussion of the countryside as offering spaces of escape away from work, study and parenting. Thus the student group highlighted the countryside as a space of escape:

> F: Yes, it is a much more relaxed atmosphere isn't it? You've not got the worries about anything else that is going on around you, you can talk about a lot more things than you would normally.
>
> F: It's a more sort of open setting, you become freer, you've got no sort of frustrations.

F: It definitely works for me, I don't know about the rest of you, but I share a house when I'm up here and it's very difficult to get time by yourself. During the university day there is everybody, and when you are home there is everybody. So it's nice to get out quietly where you're not going to get loads and loads of people. (Students – Oxford)

In a different context, the country sports enthusiasts commented on the local countryside as providing much-needed relaxation from the pressures of work:

M: I go out there, I have quite a high pressure job, I'm with the public all the time which aren't the easiest people to [be with], it's not the easiest environment to be in, and I find just being alone with my dogs out on the shore, I can think my own thoughts, I can be me, and it just flows . . .

F: I mean I'm really busy and my life is chaotic and when I go out it's peaceful, I can put things into perspective. If I've got a problem I can just walk and they just tend to solve themselves . . .

F: It does, it puts things into perspective, and it's nice when you're busy, and having to work on people all day and chat to people all day, just to get away. (Country sports enthusiasts – Arnside)

In these discussions, enjoyment of woods did not feature as a specific self-contained experience. When people described woods, they presented them as settings for particular embodied activities such as an evening stroll, cycling or playing with their children. Such activities were discussed, too, in conjunction with other outdoor settings, such as the sea and rivers, mountains and moorland, and open farmland. What was important was rarely related to the intrinsic character of the outdoor spaces themselves, but rather to the human experiences and social relationships that such spaces afford. There were many accounts of the recreational and even therapeutic power of the outdoors to afford people spaces to relax and unwind, to spend time on one's own away from habitual stresses and other places, to talk more freely and directly with friends and loved ones, to let the body be at home in itself.

For some people this was best achieved by spending time by themselves, where they valued an unmediated bodily relationship with nature. This is shown below:

Mod: When you're by yourself, what's it like?

F: I think it's a completely different frame of mind really. I mean, if you're with other people you are always in the slightly verbal sort of state, but if you're just on your own and you're sitting on a bank of a river, you're just sort of listening really.

All: [General agreement]

F: You haven't got to make that effort to respond to other people and be thinking on their wave-length, you can just be on your own wave-length and be receptive to whatever's out there.

Mod: Roy mentioned spiritual, does that resonate with anyone else?

F: Mmmm, I think when you do cut out all the kind of vocal and the social, and you are just actually sitting, the river is your companion, as it were.

M: You can use all your senses, your sense of smell, sense of touch, your sight, colours everywhere. (Outdoor enthusiasts – Lancaster)

This quote also brings out well the close embodied sensuous relationship with nature that can be afforded. People can use all the senses directly, without other people intruding or mediating that experience. This can also be found even where the walking is in the apparent company of others. The following is a description of the intimate friendship between two men, and how this evolved through the shared, often silent, bodily proximity of walking side by side:

M: There's many a day when I've been on the hill with my companion, and spoken in six hours perhaps six minutes. You are just with your own thoughts. You're preparing for the day ahead. You can do that quietly, and you're enjoying the sunshine, enjoying the countryside.

Mod: Even though you don't talk to your companion, is it still important that you have a companion with you?

M: Oh yes, oh yes. [In what way?] Well you've got camaraderie, and true friendship, that's when silence is not awkward. Difficult to interpret that. But you can meet a person for the first time, as we are [in the focus group], we're now getting to know each other, that's good. And for the first ten minutes if you'd noticed, there's this kind of frigidity as it were, and we were not terribly happy with each other. By the end of the afternoon we will be quite content, quite happy, I hope. And over a number of years, when you've got a constant companion, then you get to know the moods of each other. And you know when there's a time to talk, and a time not to talk.

Mod: And does being in the open air contribute towards that sense?

M: I think it must, yes. (Retired – Bridge of Allen)

A similar friendship was described by two ex-miners who were long-established walking companions. Walking in the fresh air had a particular resonance for these men who described the mining tradition of walking in the mountains and of their own deeply held desire for the affordances of fresh air and open spaces. As one of them said, 'we have been down the dark hole all day, so you want to get up there'. For these men, laid off prematurely and with little hope of employment, walking provides a welcome physical activity in which to structure the routines of daily life, somewhat analogous to the intense bodily physicality of mining.

Many people's desire for the outdoors, particularly those seeking tranquillity and relaxation, depends upon accessibility to spaces that they see as free from signs of human interference and control, that are other to the city. What many people desired was an unmediated relationship, in which they experience a profound engagement with oneself or others through a 'raw' and unmediated nature. Such experiences tend to involve less formal activities and uses, usually involving walking under and through trees, woods and forests.

Woods, Forests and Trees

Woods and trees are seen as affording particular settings for relaxation, tranquillity and bodily relaxation. They are spaces in which to relate to and appreciate nature and as providing particular opportunities for play, especially where those woods are composed of old broad-leaved trees. This desire for tranquillity was very much part of the experiences of fathers:

M: It's relaxing . . .
M: It also takes your stress away from your working, it takes you off your work completely . . .
M: I can walk for miles through a wood.
Mod: So what is it specifically about woods which makes them more relaxing?
M: Woods can isolate you. Even if you go into some of the city parks and they've got a small wooded area you feel isolated in it because it deadens the sound of the traffic round about you. (Fathers – Glasgow)

Many people found refreshment and revitalization when in woods. Their *therapeutic* quality was described by a woman in the outdoor enthusiasts group: 'as soon as our relationship gets under stress we need to go to the woods, and that does us, it really sorts us out'.

But what was it about trees and woods that produced these therapeutic qualities? Aside from experiencing woods as peaceful and tranquil settings where they could escape the perceived stresses of modern life, people also valued them as places in which they could value and appreciate 'nature' and wildlife. At times this encouraged the latent sense of themselves as embodied beings:

M: You walk out in the countryside and the wind's blowing you can see the gust on the trees and the trees waving. It's just [great] to see the way the fauna and the vegetation actually moves and stuff. It's like if you watch rivers, you see that the flowing of the river is very tranquil and very therapeutic . . .
Mod: So there is something quite therapeutic about trees?
F: I don't feel alone when I am around trees really [laughter]. I think trees have been there so long. . . . They seem so alive. (Students – Oxford)

Another group describes the significance of surprise within woods:

F: I think it's a mystique, you go in there and you never know what you are going to see, and you see a woodpecker or something, [and think] that's great, even though you've seen a woodpecker hundreds of times before.
F: It's always a surprise.
F: I long for the trees to get their leaves again . . . when the leaves come back in spring. Trees are so essential to my well-being. . . . That's the planet that we live on, with trees on it, and it's meant to have [them], and we're meant to be amongst them. That's the only way I can put it. (Outdoor specialists – Lancaster)

Some people also expressed a desire to plant trees because this would help re-establish a more natural (i.e. original) environment. One of the 'outdoor enthusiasts' reflected that, on an evolutionary time-scale, it was only recently that wooded areas had been cut down and wiped out. Six or seven hundred years ago it was all woods, an observation that reinforced his assertion that 'we're almost designed to be in woods' since for him woods are a 'natural' habitat for humans.

Other groups focused rather more on the specific sensuous qualities of woods, on the colours, seasons, smells, textures, shapes, rhythms and sounds associated with favourite trees and woods. These embodied sentiments tend to be associated with people's earliest memory of a wood, which provide important bases for childhood memories:

F: Mine was a bluebell wood. I was eight, I can remember going through a bluebell wood that was just carpeted and it just seemed to go on and on and on and it was just this fabulous blue. It was really, really lovely and I have always remembered that.
Mod: Was it the colour you remember?
F: It was the scent, the fragrance, it was warm and the fragrance tended to just hang and you could just smell it. That and the colour, blues and the greens and the bright lime greens of spring, the fresh breeze of spring. (Country sports enthusiast – Arnside)

Various groups spoke of the contribution of the different bodily senses, particularly smell, in their earliest childhood memories of woods, of playing in woods, of getting lost, making dens and of generally creating a self-contained world away from adults and parents. Particular memories mentioned included the smell of dampness, everything looking wilder, woollier and darker, a photo of oneself in a poncho sitting on a log with an enchanted expression, swinging in trees, listening to nursery rhymes on a tree stump, following trails made by one's father with the coloured leaves of autumn, being chased into a wood by geese, walking with one's mother to collect toadstools, and water running through rock piles. The immediacy and intimacy of smell seem particularly important in evoking memories of specific woodland places (see Tuan, 1993).

For the mothers' group, these childhood memories tended to engender a sense of sadness as they demonstrated an embodied freedom and lifestyle that was possible when they were young, but which was not afforded to their children. Woods have become places of danger which require greater regulation to make them safe again:

F: I just remember picking bluebells. There was a gang of us there, always down there in the day. It was safe then . . .
F: I just spent my life in the woods. I was always in the woods. I remember this big old oak tree had fallen down and we made a den in it, its branches had fallen across the steam and that was our bridge across to the other side.
F: There was a cave in the woods where we used to go to, and we used to say it was the dragon's cave and we used to scare people that came near. We used to just play and play . . .
F: When I was about five my auntie used to take us down to woods where she and my father were born and show us where the houses had been. And we used to go and pick blackberries and fish for tadpoles. . . . So we learnt a lot about my father's childhood . . .
Mod: Is it still possible do these kind of things now?
F: I went back down where I used to play about a year ago. The woods are still there but it's all barbed-wired off; you cannot get in there. (Mothers – Cardiff)

The vividness of such childhood memories of woods also features within cultural myths, literatures and poems. There were shared and long-established cultural repertoires that people drew upon in describing their memories: the deep forest as inspired by Hansel and Gretel and Babes in the Wood, playing Robin Hood, swinging from trees like Tarzan, finding woods just as Enid Blyton would have described them and so on. These associations may partly explain why most people expressed a strong desire for ancient broad-leaved trees, as symbols of freedom and liberty and, indeed, as the very antithesis of the 'regimental' and 'serried ranks' of commercial conifer plantations. The specific embodiment of Englishness is conveyed below in a passage in which one of the students is describing the sense of change that he particularly likes about trees and mixed and deciduous woods:

> M: It's one of the joys of the change of seasons, so that when you go out for a walk you don't really see the same things every time. If you go out for a walk every month or something then you see different things, there's different fauna and vegetation comes up during the year, whereas if you go through the town you'll find that everything virtually stays the same apart from the light coming down from above. It's the actual changing of the woodland and the environment that makes it so English, it continually changes, rather than just one monotonous view if you go through built-up or urban [areas]. (Student – Oxford)

The supposed 'Englishness' may partly explain why inner-city Asian youth are not attracted to woods. By contrast with the other groups, they expressed little desire to experience nature in the raw, claiming to enjoy woods and forests only insofar as they enabled them to participate in adventure sports. This detachment was evident, too, in their childhood memories of woods and trees. For example, two men remembered feeling pleased when trees were cut down in their garden since this meant that leaves would not create an 'autumn mess' that blocked the drains. More generally, these young men and women felt little sense of belonging within the countryside, preferring more 'civilized' activities in the cities. As one young man said:

> M: I'm not much for a countryside; peace and tranquillity is not me, I don't like it. I've been back to Kashmir a few times, well, twice now, and I get bored of it quite quickly. I prefer the city life kind of thing. (Asian youth – Spark Hill)

Although their detachment partly stemmed from the practical difficulties of getting out into the wider countryside and from a diffuse sense of ethnic difference and embarrassment in rural areas in Britain, it was also warranted by the different cultural associations of woods and forests (see Taylor, 1994, on representing the [white] English countryside). For them, trees are more likely to be thought of as useful sources of firewood. And for the young women, woods are thought of as dark, dirty and scary places, places of bodily threat. This was also evident in their idealized images of trees. This was the only group who preferred dense conifer plantations which they described as 'peaceful', 'warm', 'not artificial' and fitting for 'nature conservation'.

Woods and forests appeared to have a different cultural resonance in Kashmir and Pakistan and did not conjure up notions of bodily freedom, escape or adventure. Interestingly, this group was the only one to discuss the global ecological need for more woods and forests, which was seen generally as a greater priority than for going for walks and personal forms of pleasure in woods:

> M: Clean the air. . . . It's the nature thing about global warming, all that stuff. That's what the countryside should be for – the conservation of nature.
> M: . . . That's the only motive that I have for more forestry.
> Mod: So what is it that you're not too bothered about . . .?
> M: The walks.
> M: I like the woods for nature. The forests here don't make all that much difference anyway. The Amazon has the most effect on that. So I just think it's for nature and for clean air. (Asian youth – Spark Hill)

Aside from the Asian youth group, the strong feelings and associations of trees and woods as embedded in many people's childhood memories account for people's expressed need for woods as places of bodily play. Building on their own childhood experiences of making dens, of climbing and playing games, especially the fathers were keen that their children should have similar opportunities (and they emphasized less the dangers of contemporary woods and the need for regulation):

> M: The kids, you can let them loose [in woods] and you can have a good laugh with them. They just run up, try to climb trees, you know, everything they see that's high they want to climb it. . . . Sometimes [you] take the camera, the video camera, and you sit and watch it, the next couple of weeks when you get it developed and you go, 'God Almighty', we're in fits of laughing, you know. 'Did you do that', 'Aye', you know. You'll join in, she'll take the kids . . . they've got up the tree, you try and get up, you just fall and you feel a right clown. A kid can do that, I canna. . . . [They] enjoy themselves, just break out, you can sit there and relax and let them do their thing. (Father – Glasgow)

Some further distinctions can be seen in the ways in which different social groups use woods and forests. Much activity in woods is informal and locally based, often in the form of regular strolls to nearby woods with one's dogs or one's (walking) partner, as part of the rhythm of daily life. For others, walking in woods is more periodic, something done at week-ends in the form of specifically organized trips and outings. For the country sports enthusiasts, the retired group, the students and the fathers, there is apparently little physical constraint upon their ability to use woodlands. For these people there are woods in close proximity or within easy access, and regular walks to woods are easy and straightforward. The retired group talked about being spoilt for choice. Woods and the wider countryside are right on their doorstep, their main constraint being their bodily health and physical well-being. The sense of contentment is conveyed below, partly reflecting their deep attachment to their locality, and their particular life-stage:

F: Keep them [woods], definitely, don't lose them.
F: Keep the wild forests here.
F: The woods here are so accessible and so convenient that we don't really need anything else.
M: I think the group here are so familiar with the woods that no-one of us are surprised, there's nothing beyond that I don't think.
Mod: You're totally at peace and happy people here!
F: You have to stay here to realize. (Retired – Bridge of Allen)

A sense of danger within woods is felt by many women, especially those living in city locations (see Burgess, 1995). Most of the Cardiff mothers felt too scared to venture into woods by themselves, while for many of the other women there was a similar, if more diffuse, sense of unease in walking alone, particularly at night, unless accompanied by a dog. In the passage below the mothers describe this sense of danger and how the presence of rangers and other signs of disciplining and organized activity would make them more secure:

Mod: Do you go into woods and forests?
F: No, not by myself I don't.
F: I spent all my childhood in the woods, but not now.
F: Too scared today.
F: I'd go if there was a gang of us, I wouldn't go by myself . . .
F: Yeah, you feel unsafe, you're not safe walking the streets too, but . . .
F: It's secluded there as well.
Mod: Would you go with your kids?
F: Yeah, yeah.
Mod: So what kind of wood would you go to?
F: Forestry Commission mostly. [Why?] Well, where there's rangers, where there's picnic benches, where there's nature trails and recognized, designated routes. (Mothers – Cardiff)

Younger people in the groups – the students, the Asian youth, the mothers and the fathers – expressed the desire to use woods in more bodily active and organized ways. For the students such activities included the physically risky, such as paint-balling, horse-riding, rock-climbing, camping and mountain-biking. There is a strong desire for organized and risky adventure and fun, and for these activities the intrinsic qualities of the wood per se matter less than the affordances of these particular activities.

The outdoor specialists and outdoor enthusiasts conceive of the affordances of woods in terms of the edibility and medicinal qualities of woodland flora such as mushrooms and herbs, and the acquisition of skills of coppicing, trapping and charcoal-making. These activities were sometimes described as part of an alternative lifestyle, where woods provided the setting for a supposedly more balanced and harmonious relationship with nature. Such aspirations drew upon traditional patterns of dwelling-with-nature but were expressed in terms of an environmentalist ethics:

> Mod: So are there any things you would particularly like to be doing more of?
> M: I'd like to learn more about wild food. I know some things I can get from the woods and eat but I'd love to go out with somebody who knew their stuff about mushrooms.
> F: Yeah me too.
> F: Yeah definitely.
> F: I'd like to know more, just generally, not just the edible plants but you know, all of them really. You know if I went out for a few days on my own I'd [like to] know what I could use, and what I couldn't use, and how I could kind of survive. Yeah, how to use it without upsetting the pattern of the wood. (Outdoor enthusiasts – Lancaster)

Other people expressed similar desires for being more bodily self-sufficient with nature through transforming their relationship with woods, with resisting dominant conceptions. Some people desired to live off the forest while in a Mongolian tent. Often accompanying such uses are desires for more affordances for 'wild camping' as opposed to the organized camp sites provided by at least some landowners such as the Forestry Commission; 'I'm constrained when it comes to wild camping. Getting away from organized camp sites but that is definitely a no-no' (Outdoor enthusiasts). More generally, affordances to be able to dwell-within-woods were interestingly expressed in the following, which draws out the difficulty that such resistant bodies would be of 'no fixed abode':

> F: I think I would like to be able to live outside. And just to make a general point about the planning laws in this country which make it illegal to live in anything that isn't a house basically. And I think that if a lot more people, especially young people could live outside – not in motor vehicles, I'm not keen on them – but you know, in various sorts of tents or home-made dwellings, especially young people with young families, so more children were growing up in contact with nature, then whatever they do with the rest of their lives they will at least have that basic sense of respect for the land . . . (Outdoor specialists – Lancaster)

The use of woods for country sports reflects a further kind of relationship that people articulate with nature. The country sports enthusiasts hunted, fished or shot, and some engaged in all three. These men and women lived in a rural village and have developed what they described as a deep relationship with that place over many years. They extolled the virtues of the *local* countryside and presented country sports as central to their bodily relationship with a kind of local nature. The passage below articulates people's embodied relationship with nature that is bound up from wild fowling:

M: Wild fowling is very rough, and very cold.... And definitely not for the gentleman shooter or anybody. It's very cold out there but being out on the marsh in the morning, seeing everything wake up, it's got to be one of the most beautiful experiences. I mean unless you've been there when you're freezing cold and you are cuddling up to your labrador because, even though it smells, it's warm. But the experience is magical. (County sports enthusiasts – Arnside)

This group, with its close and intimate relationship with the countryside, adopted a somewhat exclusive understanding of who, and what, should be in that place. Some were dismissive of visitors and tourists to the area, who, it is believed, understand little of country ways and fail to comprehend the rhythms and taskscapes of country life, especially as reflected within these local sites (this view has fed into support for the Countryside Alliance):

F: Well look what's happened now they have opened up Warton Crag. Warton Crag really used to be quiet and peaceful and lovely to walk on and now it's been opened up. It's been opened up and it's been advertised and there are swarms of people that go over it, and they in themselves are causing its decline.
F: Well they are destroying the very thing they have come to see.
Mod: So you are saying the problem is the sheer volume of people?
F: Definitely happens in a wood because as soon as you have people tramping over a wood out of control, then you don't have any saplings coming up, you know.
F: They don't realize that they are living, working woods and people have to work in them and do things. It's not just a pretty little wood where we go out on a Saturday and walk through. (Country sports enthusiasts – Arnside)

Finally, when discussing the Asian youth group, we noted their preference for conifer plantations. All other groups were critical of the affordances provided within such 'forests'. The ex-miners at Merthyr Tydfil, who had had long experience of Forestry Commission woodlands through conifer plantations in local valleys, were especially disparaging about them, and had even invented the term 'a forestry' as a short-hand for such plantations. The passage below reflects how ex-miners used this distinction to demarcate the bodily affordances of (broad-leaved) woods from those of (conifer) forests:

Mod: Do you go into woods and forests much?
M: I don't ever, it's too dense in there, isn't it . . .
M: Nothing in there to see. . . . They're so close, the trees are so close, pointless going in there.
M: Can't go through forestry. Woods is different, the only woods I know up here are by Aberdare Tennar . . .
Mod: So, how is a wood different from a forestry?
M: Oh, it's different to a forestry because you can walk in it.
M: It's more open, the trees, the sky . . .
M: In the wood there's different types of trees, isn't there. Forestry is all the same.
M: The woods is wild isn't it . . .

> M: What we used to call the woods as youngsters, they've all been wiped out and there's buildings gone on them. (Ex-Miners – Merthyr Tydfil)

For these men, the Forestry Commission is a remote body that has planted dense, uniform trees in the valleys, fenced them in with little consultation with local people, and in the process removed existing paths and rights of access. There was little connection seen between such 'forestries' and the woods that the men remembered from their childhoods, deciduous woods full of bodily affordances. Various groups also argued that the principles of commercial forestry were anathema to the principles of a living, working, English landscape. A country sports enthusiast in Arnside argued:

> I want my grandchildren to see the trees that I have seen, to see the beauty of the colours changing, not to see a Norwegian landscape, which is what we see in a lot of places we go to. I want them to see an English landscape.

However, we have seen that unless such woods provide appropriate affordances then many other social groups are in effect excluded and only some will be able to immerse their bodies within such English spaces.

Conclusion

We have seen that the experience of trees and woods appears to have intimate bodily significance for many people – as providing contact with nature, as a source of tranquillity, and as a distinct 'social' space where people feel at one with themselves and their family or friends. Such affordances are mainly associated with broad-leaves and 'informal' multi-species 'natural' woods, rather than with 'commercial' forestry plantations. Whereas the vital and self-generating qualities of deciduous woods afford an active and corporeal body, the temporary plantations of trees involved in commercial forestry tend to deaden the body and make it passive (see Rackham, 1986 for a more in-depth account of the distinctions between woods and forests). Further, the affordances of woods tend to be *local* in that specific 'local' circumstances and experiences shape people's sense as to what is necessary or desirable for their bodily engagement with such spaces.

For some groups trees *are* nature – as live, sensuous indicators of the changing seasons, offering an intimate connection with seasonal cycles and deeper senses of time, through growth and regeneration, changing colours, natural variety, smell, and experiences of tranquillity, peace and the mysterious. Indeed, when people speak about trees and woods, they appeared to be trying to articulate a personal sense of what they mean by *nature*. Such sentiments are underpinned in many cases by potent childhood memories of happy woodland play, and are in sharp contrast to the perceived pressures of everyday working or domestic routines and especially life in cities.

We have seen from the different focus groups that the bodily affordances that people gain from woods and forests are influenced by personal and family life-stage, socioeconomic circumstance and geographical location. The value people appear to attach to woods and forests arises from the specific 'affordances' that the latter could offer for particular bodily desires. Thus young mothers want secure, wardened play spaces for families and children. Fathers with teenage children want open-air facilities for strenuous bodily outdoor pursuits (preferably in appropriately spectacular settings). Retired and/or unemployed country-dwellers want locally accessible open spaces for walking that are full of surprises with a lack of regulated space. Adventurous younger people desire places for risky adventures or for camping 'away from it all', and especially away from parental regulation. Field sports enthusiasts seek to maintain local, relatively undisturbed places for their sports without there being too many intruders. Asian youth mostly seek to avoid English woods altogether. Owners of woods and forests have increasingly to manage these conflicting desires for very different affordances. It is clear that, so far, this has not been managed in ways that many of the different social groups find appropriate.

In particular we found a plea for greater bodily involvement in woods and forests. This appeared to transcend the conventional distinctions between 'recreation', 'environment' and 'education'. This was because what many people appeared to be articulating was a desire for deeper, more continuously *engaged* connections with the life and continuities of woodlands in their areas. Such patterns of relationship are not captured adequately either by the conventionally 'objective' idioms of science-based conservation and 'biodiversity', to which most public bodies or commercial organizations tend to be attuned, or by the technical vocabularies of 'recreation' professionals.

In the Introduction to this volume, we talk of the view that there are 'unmediated experiences' of nature, that are not so much narratives of humans saving nature as 'nature saving us'. In this research on the leisure landscapes of woods and forests we found many groups whose experience of being outdoors in wooded nature provided exactly this sense of being saved. People eloquently talked of the different, more bodily relaxed ways of relating to one another that are afforded by being outdoors and in the woods. This closeness under the trees does not necessarily involve communication through conversation. It is rather that being close in the open air, and passing through and under the complex intertwinings of roots, trunks and branches of the mixed woods of 'nature' involves a complex resistance to the 'others' of work, study, domestic labour, the city, the modern.

Notes

We are especially grateful for the comments of Robin Grove-White, Greg Myers and Mark Toogood on earlier drafts of this article. The research reported here was funded by the Forestry Commission, took place early in 1998 and is written up in detail in Macnaghten et al. (1998). Phil Macnaghten especially thanks his fellow researchers Robin Grove-White, Claire Waterton and Sue Weldon.

1. Existing research especially by both the Forestry and Countryside Commission (see Bellringer and Gillam, 1994, 1995, 1996; Burgess, 1993, 1995; Forestry Commission, 1998; Gillam, 1997) does not provide analyses of the embodied character of people's relationships with trees.
2. These groups include a group of students from Oxford, Asian youth from Spark Hill in Birmingham, ex-miners from Merthyr Tydfil in Wales, mothers from Cardiff, retired men and women from Bridge of Allen in Scotland, fathers from Glasgow, country sports enthusiasts from Arnside in Lancashire, outdoor enthusiasts from Lancaster and people with outdoor specialist interests also from Lancaster. Each group consisted of seven to nine participants and lasted approximately two hours. The sessions began with a wide-ranging discussion of people's experiences of 'the outdoors' and of their sense of bodily opportunity and constraint. People were then encouraged to discuss their feelings, memories and bodily experiences of woods and forests; their favourite kinds of woods and trees and why; the range of preferred uses and activities; and their aspirations for trees, woods and forests in the future. To enable them to explore and discuss their views, photographs of diverse woodlands were introduced as stimulus material.

References

Bellringer, A. and S. Gillam (1994) *Forest Visitor Survey Summaries*. Edinburgh: Forestry Commission.

Bellringer, A. and S. Gillam (1995) *Forest Visitor Survey Summaries*. Edinburgh: Forestry Commission.

Bellringer, A. and S. Gillam (1996) *Forest Visitor Survey Summaries*. Edinburgh: Forestry Commission.

Bonnemere, P. (1998) 'Trees and People: Some Vital Links', pp. 113–31 in L. Rival (ed.) *The Social Life of Trees: Anthropological Perspectives on Tree Symbolism*. Oxford: Berg.

Burgess, J. (1993) *Perceptions of Risk in Recreational Woodlands in the Urban Fringe*. Cheltenham: Countryside Commission.

Burgess, J. (1995) *Growing in Confidence: Understanding People's Perceptions of Urban Fringe Woodlands*. Cheltenham: Countryside Commission.

Clark, G., J. Darrall, R. Grove-White, P. Macnaghten and J. Urry (1994) *Leisure Landscapes*. London: CPRE.

Costall, A. (1995) 'Socialising Affordances', *Theory and Psychology* 5: 467–81.

Forestry Commission (1998) *United Kingdom Visitor Survey*. Edinburgh: Forestry Commission.

Giambelli, R. (1998) 'The Coconut, the Body and the Human Being: Metaphors of Life and Growth in Nusa Penida and Bali', pp. 133–57 in L. Rival (ed.) *The Social Life of Trees: Anthropological Perspectives on Tree Symbolism*. Oxford: Berg.

Gibson, J. (1979) *An Ecological Approach to Visual Perception*. Boston, MA: Houghton Mifflin.

Gillam, S. (1997) *Public Opinion of Forestry 1997*. Edinburgh: Forestry Commission.

Henwood, K. and N. Pigeon (1998) *The Place of Forestry in Modern Welsh Culture and Life*. Bangor: University of Wales.

Ingold, T. (1993) 'The Temporality of the Landscape', *World Archaeology* 25: 152–74.

Knight, J. (1998) 'The Second Life of Trees: Family Forestry in Upland Japan', pp. 197–218 in L. Rival (ed.) *The Social Life of Trees: Anthropological Perspectives on Tree Symbolism*. Oxford: Berg.

Macnaghten, P. and J. Urry (1998) *Contested Natures*. London: Sage.

Macnaghten, P., R. Grove-White, C. Waterton and S. Weldon (1998) *Woodland Sensibilities: Recreational Uses of Woods and Forests in Contemporary Britain*. Lancaster: CSEC.

Rackham, O. (1986) *The History of the Countryside*. London: J.M. Dent.

Rival, L. (ed.) (1998) *The Social Life of Trees: Anthropological Perspectives on Tree Symbolism*. Oxford: Berg.

Schama, S. (1995) *Landscape and Memory*. London: HarperCollins.

Schiva, V. (1989) *Staying Alive*. London: Zed.

Taylor, J. (1994) *A Dream of England*. Manchester: Manchester University Press.

Tuan, Y.-F. (1993) *Passing Strange and Wonderful*. Washington, DC: Island Press.

Uchiyamada, Y. (1998) '"The Grove is Our Temple": Contested Representations of Kaavu in Kerala, South India', pp. 177–96 in L. Rival (ed.) *The Social Life of Trees: Anthropological Perspectives on Tree Symbolism*. Oxford: Berg.

Zelter, A. (1998) 'Grassroots Campaigning for the World's Forests', pp. 221–32 in L. Rival (ed.) *The Social Life of Trees: Anthropological Perspectives on Tree Symbolism*. Oxford: Berg.

Phil Macnaghten is Lecturer in Culture and Environment at the Centre for the Study of Environmental Change, Lancaster University. His research interests centre on the relationship between environmental concerns and wider societal trends. Recent research has looked at public controversies around genetically modified foods, the millennium, forestry, sustainability and new politics of the environment. His publications include *Contested Natures* (with John Urry; Sage, 1998).

John Urry is Professor of Sociology at Lancaster University. His recent publications include *Economies of Signs and Space* (with Scott Lash; Sage, 1994), *Consuming Places* (Routledge, 1995), *Touring Cultures* (co-edited with Chris Rojek; Routledge, 1997), *Contested Natures* (with Phil Macnaghten; Sage, 1998) and *Sociology Beyond Societies* (Routledge, 2000).

Perceiving the Environment in Finnish Lapland

TIM INGOLD AND TERHI KURTTILA

On the Meaning of Traditional Knowledge

In a recent article, anthropologist Bjorn Bjerkli reports on his investigations into the traditional pattern of using common land among a group of farmers, mostly Sami-speaking, inhabiting a small valley not far from the city of Tromso, in northern Norway (Bjerkli, 1996). He is concerned because this pattern has recently brought them into conflict with the authority of the Norwegian state, which asserts that they have no right to use the land in the way they do. The case is about to come to court. The difficulty is that in order to justify in law their claim to use the land as a commons, local people will have to demonstrate that this use has been subject, from time immemorial, to clearly articulated (if heretofore unwritten) rules and procedures. So far they have been unable to do this, and have only come up with very vague statements about 'the way we do things here'. They can talk about what went on in the past, about people and events, good years and bad years, and so on. But they cannot formulate explicit principles of 'traditional' land use. Bjerkli is worried that if they are forced to enunciate such principles in order to win their claim for land, then in future they will be required to regulate their land use according to these principles. And the effect of this regulation could be to destroy the very tradition they seek to sustain.

The crux of the problem lies in the concept of 'tradition', to which lawyers, bureaucrats and politicians appeal just as often and freely as do local people. For it is clear that they mean different things by it. Below, we suggest what some of these differences are. First, however, it is necessary to reflect a little on our own academic practice. For in much of the discussion about so-called 'traditional

Body & Society © 2000 SAGE Publications (London, Thousand Oaks and New Delhi),
Vol. 6(3–4): 183–196
[1357–034X(200009/12)6:3–4;183–196;015477]

environmental knowledge', in anthropology and related disciplines, the sense in which the concept of tradition is used comes closer to that adopted by the state administration than to that understood by local people. One solution might be to drop the concept of tradition altogether: to regard it as so tainted by its conventional opposition to modernity that it can give only a distorted view of people's real lives, one that is flattened in time and devoid of any sense of history.[1] But this solution is hardly satisfactory in a situation where local people value, and want to continue with, what they themselves see as a traditional form of life, and where they have to contend with administrative authorities for which 'tradition' is taken to be an essential ingredient in the management of local affairs. A better solution, we think, is to find a way of talking about tradition that chimes more accurately with local sensibilities, and to reconstruct our theory of 'traditional ecological knowledge' around this. In what follows we make some attempts in this direction.

Our principal objective is to establish a contrast between two understandings of traditional knowledge, one embedded in the modernist frameworks of the state apparatus, the other in the everyday life of local people. For simplicity, we shall call these MTK (traditional knowledge as enframed in the discourse of modernity) and LTK (traditional knowledge as generated in the practices of locality) respectively. As the rather circumlocutory form of the expressions placed in parentheses reveals, we have had to choose our terms with some care. The literature is replete with dichotomies of one kind and another, all of which aim to adumbrate the particular character of the local, the non-Western and the traditional against the background of a uniform and generally characterless global, Western modernity. Our intention is not to propose yet another opposition of this kind. It is rather to show that the terms in which such oppositions are constructed are fundamentally incommensurable with the lived experience of the people with whose knowledge and practices we are concerned. Thus we contend that the construction of 'indigenous knowledge' (MTK), through its opposition to modern science, implies a sense of what it means to know that *dis*-places the knower, and that is incompatible with ways of knowing (LTK) that are constitutive of locality.

It is important to warn against another possible misapprehension of the nature of the contrast we are drawing here, not least because it is one to which, initially, we fell victim ourselves! In an earlier version of this article we defined MTK and LTK as traditional knowledge, respectively, in 'modernist conception' and in 'local conception'. This made it appear as though people educated in the mainstream of Western modernity shared one concept of tradition, whereas local people shared another. The task, then, would be to compare 'our' concept with 'theirs'. Yet as we shall show in a moment, the idea that people's experience is organized in terms of shared concepts that are transmitted through their education in a particular culture belongs, fair and square, within the modernist discourse on traditional knowledge. To compare LTK and MTK in such terms would be to confirm rather than challenge the hegemony of this discourse. Our aim is different. Far from engaging in an anodyne exercise of cross-cultural comparison, we have set out to destabilize the epistemological foundations upon which all such comparisons are built. We propose that the meaning of tradition, for local people, is not itself handed on as part of a received 'cultural model' for the interpretation of experience, but is drawn from those interactive contexts in which people become consciously aware of their own particular knowledgeability, a knowledgeability that has its source in the very activities, of *inhabiting* the land, that both bring places into being and constitute persons as *of* those places, as local.

Now one of the characteristic things about MTK is that it is very closely bound up with what we call the 'genealogical model'. This is based on the idea that the rudiments of make-up and identity that go together to constitute a person are *received*, along one or several lines of descent, from that person's ancestors, and will in turn be *passed on* to his or her descendants. It is commonly supposed that these attributes are of two kinds, biogenetic and cultural, the former carried across the generations along with components of bodily substance, such as 'blood' or 'genes', the latter passed on through a somewhat analogous transmission of rules and representations, coded in speech or other symbolic media. These rules and representations add up to what is known, in the language of the genealogical model, as the 'cultural heritage'. However, if culture is thus transmitted along lines of descent, then its essential content must be specifiable independently of the actual environmental circumstances that its individual recipients encounter in the course of their lives. For any particular person, the acquisition of traditional knowledge – through some process of enculturation or social learning – is assumed to take place both independently of, and prior to, its application or expression in real-life contexts of activity. This assumption, which isolates the intergenerational transmission of knowledge from environmentally situated experience, lies at the core of the genealogical model.

That the significance of this model, and its impact, extends far beyond the confines of academic anthropology is evident from much of the controversy surrounding the contemporary political and legal status of the world's so-called 'indigenous peoples'. According to the 1989 Convention of the International Labour Organization, people are regarded as indigenous 'on account of their *descent* from populations which inhabited the country . . . at the time of conquest or colonisation or the establishment of present state boundaries' (ILO, 1989:

Article 1.b; emphasis added). Thus the traditional knowledge of contemporary indigenous peoples relates not to current circumstances of using the land but to an earlier, pre-colonial era when their ancestors were its only inhabitants. Converted into an object of memory, this original connection to the land has supposedly devolved, in an unbroken chain of descent, from the ancestors to their contemporary descendants. To claim, however, that traditional knowledge is received fully formed from genealogical antecedents, as a legacy from the past, is tantamount to an admission that people's present-day experience of inhabiting the land plays no further part in its constitution. So long as the stuff of tradition can be passed along, like a relay baton, from generation to generation, it should make no difference, in principle, where the people are, whom they live with, or what they do for a living. One could indeed qualify as an indigenous person, on the criterion of descent, without once setting foot in ancestral country.

The principles of the genealogical model, built into the formal definition of indigenous status, have in their application had fateful consequences for the people so designated. For administrations have had frequent recourse to these principles to justify policies of removing indigenous populations from the land. To ensure the continuation of valuable traditional wisdom, it is argued, no more is needed than adequate institutional mechanisms for its storage and replication. Thus, resources for the preservation of indigenous cultures are put into museums, schooling in native language and handicraft, folklore research and so on. For local people, by contrast, traditional knowledge is inseparable from actual practices of inhabiting the land. For it is in the relationships that are forged with the land, along with its animal and plant life, that their knowledge is generated. Thus when the Sami farmers whom Bjerkli was studying say of their tradition that 'this is how we do it here' (Bjerkli, 1996: 9), they are referring to knowledge founded in activities of livelihood that make the land, for them, a *place*. Perhaps they might say that a really traditional person is one who knows the country 'like the back of his hand'. This does not mean that he carries it in the form of a cognitive map inside his head. But it does mean that, through having grown up there, he has learnt to 'know' it rather as an experienced craftsman might be said to know his raw material. That is, he is acutely sensitive to its forms and textures, can respond creatively to its variations, and is ever alert to the possibilities these afford – and the hazards they present – for pursuing different kinds of tasks. Such, in short, is the nature of LTK.

Climate and Weather

To give a concrete illustration of what we mean by LTK, we should now like to report briefly on the results of a recent project of research into the ways in which

Sami people living in northernmost Finland perceive their environment.[2] We were particularly concerned to find out whether, or in what respects, the environment is perceived to have changed – at least within living memory – and whether these changes could be linked to wider processes of climatic change as 'scientifically' monitored and recorded by meteorological stations in the region. If such a link could be established, we reasoned, then it would give us a way of incorporating local knowledge into the process of regional climatic modelling. And at the same time, we would have a basis for converting predictions derived from climate change models into meaningful prognoses for people at the local level. Our efforts, however, were frustrated by the realization, in the course of carrying out the project, that environmental scientists and local Sami people were talking about quite different things. In a nutshell, whereas the scientists were out to detect changes in *climate*, what mattered to local people were changes in the *weather*. Climate is an abstraction compounded from a number of variables (temperature, precipitation, air pressure, windspeed, etc.) that are isolated for purposes of measurement. Weather, by contrast, is about what it feels like to be warm or cold, drenched in rain, caught in a storm and so on. In short, climate is recorded, weather experienced. And our attempts to integrate climatic records with stories about weather-related experiences indicated that the relation between climate and weather is anything but straightforward.

Fieldwork carried out by one of us (Kurttila) revealed how experiences of the weather are closely bound up with individual life-histories: memories of childhood, stories of coping with adversity, wartime exploits and the like. Indeed, the interview material often disclosed more about processes of memory than about actual climatic episodes. Nearly always, observations concerning the weather were woven into narratives relating to the performance of particular tasks, such as reindeer-herding, berry-picking or fishing, or to life-crisis events such as childbirth, illness or injury, and death. As such, they formed part of the ongoing construction of those familiar places, along with their surroundings, that people call 'home'. Respondents generally found it difficult to recall what the weather was like during periods when they were away from home (for example, attending boarding school). Thus the work of memory, and hence people's sense of continuity with their own past, was intimately tied to their experience of inhabiting particular locales. Then again, older people, judging recent experience against a long-remembered past, often had ideas of seasonal normality that differed from those of the younger generation. Some old people were credited with exceptional ability in reading the signs, in the environment, of impending weather changes. But such knowledge of the weather is not something that is handed down as a set of customary prescriptions or formulae; rather it grows through a lifetime's experience of living in a place and moving in its environs.

Our research also revealed a close connection between the ways people perceive the weather and both their tasks of work and modes of travel. Because men and women often engage in different tasks, their perceptions differ accordingly. Men move more widely in the forest, rounding up reindeer or hunting, and are sensitive to the ways in which warmth or chill, rain or snowfall, and especially the wind, affect their own movements and those of forest creatures. Women spend more time about the house, and are mindful of the weather insofar as it impacts on domestic chores, the growth of garden vegetables such as potatoes, the well-being of children (making sure they are properly clothed) and so on. Women do venture widely into the forest in season, to gather berries, but one's experience of the weather when crouched on the forest floor is very different from when one is running after reindeer! There are task-related differences in the perception of weather, too, between adults and children. The regular journey to school, nowadays normally undertaken by bus, figures prominently in children's experience. But so do the joys of bathing in the lake in summer to keep cool and escape the mosquitoes, of skating on the ice in winter and of being able to ride a bicycle on the hard snow-crust in spring. 'I was able to cycle to school over the snow', as one man recalled, 'and of course I was always late. It was such fun that I made a lot of extra circles there in the forest.'

Since perceptions of weather depend on tasks and travel, changes in working practices, and above all in transport technology, have led to new ways of perceiving weather phenomena. For example, the adoption of the snowmobile for winter travel calls for a quite different appreciation of the depth and consistency of snow, while the widespread use of motorbikes and four-wheel 'crawlers' on open ground in summer affects the perception of the land surface. Just as the ground feels different under wheels from the way it feels underfoot, so also the feeling of snow under the tracks of the snowmobile is different from that under skis. With snowmobile travel in particular, any failure to judge surface conditions correctly can carry very real dangers for machine and driver alike, nowhere more than in travelling over snow-covered ice. The speed of motorized travel further magnifies the impact of wind and the severity of wind-chill. Of course the construction of the road network, and the possibilities for travel by bus or private car, not only afford a degree and speed of mobility unimaginable to earlier generations, but also insulate the traveller to the extent that the practice of travel is not so totally an experience of weather as it used to be. One can, for example, see the frost without being bitten by it, hear the rain without being drenched in it, or watch the wind in the trees without feeling it on your face. Thus people are coming to perceive the weather less through immediate bodily experience, and more in terms of how it affects the performance of their vehicles. However, as swathes of cleared and levelled ground, artificially cut through a mainly forested landscape, roads have their own weather, known to anyone whose business of life lies along them. They are windier and colder in winter, and more hot, dry and dusty in the summer.

For Sami people, the perception of weather is multisensory. It is just as much auditory, tactile and olfactory as it is visual: indeed, in all normal circumstances these sensory modalities cooperate so closely that it is quite impossible to separate out their respective contributions to the totality of weather-related experience. One perceives, in effect, with the whole body. Nevertheless, hearing is critically important, more so, probably, than any other sense. A person listens out for birdsong, snow squeaking underfoot, the crackling of a fire, human voices, distant reindeer bells, dogs barking, ice cracking, thunder and so on. These are not background sounds; they punctuate the space of lived experience rather than filling it up. Smells, on the other hand, do tend to fill the air. Some people say they can smell when it is becoming warmer on cold winter days, presaging snowfall, and of course, the forest smells quite differently after summer rain. Touch comes to the fore in the feeling of the ground under one's feet and of the wind on one's face. The weather can also be felt with the eyes: the dryness and dust of a hot summer's day, or the bite of the wind in a severe frost. While people of course use their eyes principally to see things, to pick out objects in their environment, the visual perception of the weather is above all an experience of light itself: the glare of sunlight or moonlight on the snow, the modulations of light and shade brought about by changes in cloud cover, the shifting shadows cast by the boughs of wind-blown trees, the lightning in a storm, the colours of sunrise and sunset, the transparency of water and ice, the flickering northern lights and so on.

The multisensory awareness of the environment that we have described above is the key to spatial orientation and the coordination of activity. In reindeer herding, for example, sensitivity to the movements of the wind, of deer and of other people, is critical. In the judgement of impending weather conditions, observations of the behaviour of animals, such as the arrivals and departures of migratory birds, are especially important. On occasions when everything falls still and silent, such as in fog or before a storm, people can feel completely disoriented, even in familiar surroundings. Many people had stories to tell of such occasions, which always contained the same moral: you can never take the environment for granted. All of a sudden, the most well-known places can 'flip over' and turn strange and hostile, leaving the traveller lost and bewildered. No-one is ever skilled or knowledgeable enough to be able to move in the forest with total confidence: so far as the weather is concerned, one has always to contend with a degree of uncertainty, and it is the recognition of this uncertainty that distinguishes the truly experienced woodsman. Above all, moving in an environment means 'tuning' one's own movement in response to the movements in one's surroundings – other animals, the wind and so on. Where nothing moves there is nothing to respond to; hence the feeling of disorientation.

Crucial to people's experiences of the weather in the far north is the cycle of the seasons. As a phenomenon of climate, seasonality may be registered in the form of regular annual fluctuations in temperature, precipitation and daylight hours, without regard to the lives of plants, animals or human beings. But as a phenomenon of weather, seasonality inheres in the relations between concurrent rhythms of growth and movement of plants and animals, and of human social life.[3] Here, for example, is one Sami reindeer herdsman's graphic depiction of the transition from summer, through autumn, to winter:

> In August the darkness comes like a thief, it catches you in a few weeks, the nights become cooler all too suddenly – everyone was expecting it but no-one was ready for it. The birds are leaving, disappearing with their cries across the sky. People are gathering berries and fishing, trying to carry on with the summer until they have to give up. Nature is full of red and yellow colours, the mornings are cold and bright, and it is a time for dying. Hunting begins for birds and elk, domestic reindeer are killed and there is long-awaited fresh reindeer meat to eat. For a while there are dark evenings and cold rain, but not for too long, for suddenly one realizes that there is a season to wait for, a season of snow and ice. Everyone talks about the first snow, and the first ice, and makes a mark in their calendar. Reindeer are returning to their winter grazing areas: it is winter again, dark nights, stars, moon and northern lights – skies with all the colours. Time to get back to work with the reindeer.

This passage brings out beautifully how seasonal variations are experienced as the interweaving, in a complex counterpoint, of changing harmonies of light, darkness and colour, of freezing and thawing, of cycles of life and death, of the migratory movements of birds, and of human activities of production (berrying, fishing, hunting, herding) and consumption (from eating fish to eating meat). And it shows, too, how every change of season embodies a mixture of anticipation and surprise. You know that autumn is coming, but still you try to hang on to summer to the last. And the arrival of winter always catches you out, however much it was expected.

It is consistent with this understanding of seasonality as a system of rhythmic interrelationships that in recounting their more memorable experiences of the weather, people tended to focus on rhythmic dislocations and the anomalous conjunctions that ensued. For example, in an exceptional year in which the snow-cover persisted until mid-summer, one respondent recalled the oddity of hearing the cuckoo calling in the midst of the snow. In another year people remembered a July snowfall and the sight of swarms of black mosquitoes silhouetted against the white background. And when particular migratory birds are late in arriving in spring and summer, their absence may be remarked upon. There is a widespread feeling, however, that the previously sharp divisions between the seasons are disappearing. At the same time, people admit that they are losing their skills of 'reading nature' for indications of impending weather change, and are relying more on radio and television forecasts. This is not, however, simply a matter of replacing a 'traditional' form of prediction with a 'modern' one. Traditionally, knowing what the weather was going to do was a matter of noticing, and responding to, subtle 'early warning' signs in the environment. What the forecast offers, by contrast, is prediction of a quite different kind, in the form of a hypothetical, map-like projection of a future state of affairs which has then to be interpreted in the light of information specific to the local area before it can be understood.

Tradition as Process and as Skill

In a recent article, Toni Huber and Poul Pedersen contrast modern meteorology with traditional knowledge about the weather. Though their source for the latter is ethnographic material from Tibet, their conclusions run along much the same lines as those derived from our study of the perception of weather among Sami people. Meteorological science, they write, 'represents the environment as an ensemble of global, quantified interrelationships', whereas in traditional Tibetan precept and practice, the weather pertained to an environment understood as 'a system of local, qualitative relationships' (Huber and Pedersen, 1997: 590). More significantly, they show how this latter understanding of the environment is fundamentally incompatible with modernist representations of traditional Tibetans as the custodians of an ancient cultural heritage founded on a principle of harmonious coexistence of humanity and nature. Traditional Sami have often been portrayed in the same way, as befits their status as an indigenous people. Yet the very notion of a cultural heritage, consisting in an elaborate system of inter-generationally transmitted representations, rests on the foundations of a genealogical model that separates out the acquisition of traditional knowledge from environmentally situated experience. As will be recalled from our earlier discussion of the distinction between LTK and MTK, the latter is supposed to consist of items of knowledge that are 'passed down' as objects of memory, *prior* to their retrieval and application in contexts of practice.

However, everything we have said about Sami people's perception of the weather – its embeddedness in personal life-histories of inhabiting particular places, its dependency on tasks and modes of travel, its multisensory quality, its role in spatial orientation and the coordination of activities, and its seasonal periodicity – flies in the face of such separation of the acquisition of knowledge from its application. Sami people do not so much apply their knowledge in practice as know *by way of* their practice. Thus their traditional weather-related knowledge, consisting in a sensitivity to critical signs in the environment and an intuitive understanding of what they mean for the conduct of practical tasks, is not really passed down at all – not, at least, in the way that this is envisaged within the modernist paradigm. It rather undergoes continual generation and regeneration within the contexts of people's practical engagement with significant components of the environment. But there are two sides to this engagement. It can be described, on the one hand, as 'how we do things here' and, on the other, as 'what being here feels like as we do these things'. A description of the first kind calls for an account of practical activities; a description of the second kind calls for an account of the weather. In that sense, practical activities and the weather are two sides of the same coin. That coin is LTK.

At root, the distinction between MTK and LTK hinges on the difference between thinking of tradition as a kind of *substance*, and thinking of it as a type of *process*. Basic to the modernist conception of the person are the metaphors of container and content. Equipped by nature with universal capacities, human beings are viewed as containers for the culturally variable, substantive content that specifies traditional knowledge in its diverse spheres of application.[4] Regarded as substantive mental content, it is in the nature of MTK that it should stay the same, from generation to generation. In reality, of course, it does not, but this is attributed to 'imperfections' in the transmission mechanism – somewhat analogous to the causes of genetic mutation – whereby knowledge is passed from one container to another. Some stuff may be lost; other stuff gained; overall, then, traditions change. LTK, by contrast, is not cognitive; it does not lie 'inside people's heads', as opposed to 'out there' in the environment. It is rather a process, one that is continually going on. This process is none other than that of people's practical engagement with the environment. The activity of remembering itself, upon which the continuity of tradition depends, is part and parcel of that engagement. In effect, people remember as they go along. In so doing, they follow a 'way of life' in the sense not of acting out a script received from predecessors, but of literally negotiating a path through the world.

In following this path, the important thing is that one can keep on going – that it should not come to a dead end or become caught in a loop that would consign the traveller to the endless repetition of identical trips. Indeed 'keeping the process going' may involve a good measure of creative improvisation, rather like keeping the music going in a jazz band. The crucial point here is that, regarded as a process, tradition can be continuous without taking any fixed form. There is no opposition in LTK between continuity and change. Change is simply what we observe if we sample a continuous process at a number of fixed points, separated in time. The growth of an organism, for example, is continuous, but if we compare its appearance at different times it will appear to have changed. So, too, the growth of LTK is an aspect of the growth of persons, in the contexts of their relations with one another and with the environment. Just because people are doing things differently now, compared with the way they did them at some time in the past, does not mean that there has been a rupture of tradition. What would really break the continuity, however, would be if people were forcibly constrained to replicate a fixed 'traditional' (*sensu* MTK) pattern. The effect would be similar to a gramophone needle getting stuck in the groove of a record. One could not keep the music going. Likewise any attempt to 'traditionalize the traditional', as Bjerkli so beautifully put it (1996: 18), would rupture the continuity of life.

In essence, LTK is a kind of knowledge that might be better denoted by the concept of *skill*. To explain what we mean by this, it is necessary to make three general points about skill.[5] First, skills are not properties of the individual body considered, objectively and in isolation, as the primary instrument of a received cognitive tradition. They are rather properties of the whole system of relations constituted by the presence of the agent in a richly structured environment. Thus the study of skill demands an ecological approach that situates the practitioner, right from the start, in the context of an active engagement with his or her surroundings. Second, skilled practice is not just the application of external force but involves qualities of care, judgement and dexterity. This implies that whatever practitioners do *to* things is grounded in an active, perceptual involvement *with* them, or in other words, that they watch and feel as they work. Third, skills are refractory to codification in the programmatic form of rules and representations. So it is not through the transmission of any such programmes that skills are learned, but rather through a mixture of imitation and improvisation in the settings of practice. What happens, in effect, is that people develop their own ways of doing things, but in environmental contexts structured by the presence and activities of predecessors.

It would be wrong, then, to say of LTK that it is 'cultural' rather than 'biological', or in the head rather than in the body. It is rather a property of the whole human organism-person, having emerged through the history of his or her involvement in an environment. Having recognized this, it is possible to understand local people's inability to formulate their traditional knowledge in anything other than the most vague and general terms. This fact has often been judged by outsiders, including legal experts and administrative officials, as a measure of its inadequacy or inauthenticity. But in reality, the vagueness or elusiveness of their formulations is the very source of their strength. They are, in effect, *rules of thumb*, very general notes of guidance that may be drawn on as resources for action, but that in no sense govern its course (Ingold, 1996b: 36–8; Suchman, 1987:

52). They cannot do so, since the normal environment for any kind of practice is never quite the same from one moment to the next, and the essence of dexterity lies in being so attuned to these variations as to be able, continually and fluently, to respond to them. To the extent that procedures are precisely specified, or movement 'pre-programmed', dexterity is compromised. In short, the vagueness of LTK's formulaic prescriptions is a condition for the sheer *precision* of the actual movements of knowledgeable practitioners.

We should like to conclude with three points. The first concerns the constitution of locality. What makes knowledge local? What makes people local? Knowledge is local rather than global, within the parameters of modernist discourse, not because it is tied to a place, but because it is in the heads of local people. And people are local because they happen to live in a particular place. Not only is the place fixed, within a system of global coordinates, independently of their living there, but also they do not need to live there in order to know what they do. For this knowledge is supposed to come down by descent from ancestors, rather than arising from the experience of growing up on the land. We have argued, to the contrary, that knowledge is local because it inheres in the activity, of inhabiting the land, that actually *creates* places. And in creating places, it also makes the inhabitants people *of* those places – it makes them local. People belong to the localities and environs in which they have grown up, just as much as the latter belong to them. To break the bond between persons and place along the lines of a division between cultural heritage and natural environment would be to cast tradition adrift from its generative source of meaning, leaving it as the vestige of an ancestral form of life long since overtaken by its representation as an object of memory. The effect of such displacement would be to rupture the continuity of tradition and to cut people off from their pasts.

The second point concerns the constitution of persons. We have shown how, in the discourse of modernity, traditional knowledge is linked to a genealogical model according to which every person is constituted from two substantive components, respectively cultural and biological (mental and material), passed down from predecessors. To understand what traditional knowledge means for local people, however, we have to see it as consisting in certain powers of perception and action, involving dispositions and sensibilities established in the course of a lifetime of practice and training in an environment. Here the person is conceived not as a substantive entity, but rather as a locus of growth and development within a field of relationships. And by the same token, the contribution that other people make to one's own knowledge – often represented in the idioms of kinship – is not one of substance but rather one of setting up the conditions in which growth can occur. Persons, in short, come into being not by having stuff passed on to them along lines of descent, and against the backdrop of nature, but through the immediate experience of sensory participation with human and non-human components of the dwelt-in world. And the knowledge born of this experience, though commonly dismissed as 'intuitive', must necessarily form the bedrock for any system of regulation or management that would treat the environment as the *object* of its concern.

For our third and final point, we return to the distinction between science and traditional knowledge. In the language of modernity, these kinds of knowledge are radically opposed, as indeed are the disciplines of natural science and cultural anthropology. Science purports to study nature as it really is, whereas anthropology is supposed to bring to light the cultural models in terms of which the people of this or that tradition organize and interpret their experience. This conventional division of labour, whereby scientists study the 'reality' of nature while anthropologists study its cultural construction, is, in our view, no longer tenable. After all, the weather, as experienced by Sami people, is no less 'real' than the climate recorded by scientists. In their work, scientists engage in certain environmentally emplaced, observational practices, just as much as local people do. In each case, knowledge grows out of this engagement. In short, *science is itself a form of LTK*: it is both traditional in its mode of reproduction and engendered in the practices of locality (Turnbull, 1993). If a distinction is to be drawn at all between scientific and local tradition, it lies not in the epistemological status of the knowledge itself, but in the nature of the skilled practices through which it is generated. Working under the tutelage of local people, anthropologists learn about the environment through participation in one set of practices, while scientists learn about it through their participation in another. Recognition that both cultural anthropologists and natural scientists, each in their own way, are learning about the same thing (the environment) rather than different things (culture on the one hand, nature on the other) could not only open up potentially rewarding avenues of cooperation but also allow the admission of local people as full and equal participants in the learning process.

Notes

The research on which this article is based was funded by the Economic and Social Research Council, as part of its Global Environmental Change Programme (award number L-320-25-3140). An earlier version was presented at the ESRC-GEC Workshop on Environmental Knowledge, held at the University of Keele in January 1999. We are grateful to all the participants in the Workshop for an extremely positive and helpful discussion. We also owe a debt of gratitude to the many Sami people who assisted us with such generosity in our research.

1. It is of course the refusal of real life to remain confined within the straitjacket of a static form that leads to the perception that tradition is always on the brink of disappearing.

2. Fieldwork was carried out in the three Sami communities of Angeli, Lemmenjoki and Karigasniemi, over 16 months from May 1995 to September 1996.

3. This approach to seasonality has been developed by Harris (1998) in his analysis of the rhythms of life among riverine villagers on the Amazonian floodplain.

4. There is much debate among cognitive scientists about whether there are few or many containers, or 'mental modules', per human being, and about whether – or to what extent – these containers constrain what is acceptable by way of content, but I shall not go into these issues here. See Hirschfeld and Gelman (1994) for a review of this debate.

5. These points are explored at greater length in Ingold (1996a: 178–9).

References

Bjerkli, B. (1996) 'Land Use, Traditionalism and Rights', *Acta Borealia* 13(1): 3–21.

Harris, M. (1998) 'The Rhythm of Life on the Amazonian Floodplain: Seasonality and Sociality in a Riverine Village', *Journal of the Royal Anthropological Institute* 4(1): 65–82.

Hirschfeld, L.A. and S.A. Gelman (1994) 'Towards a Topography of Mind: An Introduction to Domain Specificity', pp. 3–35 in L.A. Hirschfeld and S.A. Gelman (eds) *Mapping the Mind: Domain Specificity in Cognition and Culture*. Cambridge: Cambridge University Press.

Huber, T. and P. Pedersen (1997) 'Meteorological Knowledge and Environmental Ideas in Traditional and Modern Societies: The Case of Tibet', *Journal of the Royal Anthropological Institute* 3(3): 577–98.

ILO (International Labour Organization) (1989) 'Indigenous and Tribal Peoples in Independent Countries', Convention 169. Geneva: ILO.

Ingold, T. (1996a) 'The History and Evolution of Bodily Skills', *Ecological Psychology* 8(2): 171–82.

Ingold, T. (1996b) 'The Optimal Forager and Economic Man', pp. 25–44 in P. Descola and G. Palsson (eds) *Nature and Society: Anthropological Perspectives*. London: Routledge.

Suchman, L. (1987) *Plans and Situated Actions: The Problem of Human–Machine Communication*. Cambridge: Cambridge University Press.

Turnbull, D. (1993) 'Local Knowledge and Comparative Scientific Traditions', *Knowledge and Policy* 6: 29–54.

Tim Ingold is Professor of Social Anthropology at the University of Aberdeen, and was formerly Max Gluckman Professor of Social Anthropology at the University of Manchester (1995–9). He has carried out ethnographic research among Saami and Finnish people in Lapland, and has written extensively on comparative questions of environment, technology and social organization in the circumpolar North, as well as on evolutionary theory in anthropology, biology and history, on the role of animals in human society, and on issues in human ecology. His current research interests are in the anthropology of technology and in aspects of environmental perception. He is a Fellow of the British Academy and the Royal Society of Edinburgh, and was President for 1999 of the Anthropology and Archaeology Section of the British Association for the Advancement of Science. His publications include *The Skolt Lapps Today* (Cambridge University Press, 1976), *Hunters, Pastoralists and Ranchers* (Cambridge University Press, 1980), *Evolution and Social Life* (Cambridge University Press, 1986) and *The Appropriation of Nature* (Manchester University Press, 1986). His latest book, *The Perception of the Environment: Essays on Livelihood, Dwelling and Skill*, was published by Routledge in 2000.

Terhi Kurttila completed her Master's Degree in Cultural Anthropology at the University of Oulu, Finland, in 1993. She has lectured at the Universities of Oulu and Lapland (Rovaniemi), and has carried out fieldwork for several research projects both in Finnish Lapland and in Western Siberia. Her interests are in the culture and identity of northern reindeer herders, ethnoscience and visual anthropology.

Index